Lecture Notes in Computer Science 6290

Commenced Publication in 1973
Founding and Former Series Editors:
Gerhard Goos, Juris Hartmanis, and Jan van Leeuwen

W0192944

Marina L. Gavrilova   C.J. Kenneth Tan
François Anton (Eds.)

Transactions on Computational Science IX

Special Issue on Voronoi Diagrams
in Science and Engineering

 Springer

Editors-in-Chief

Marina L. Gavrilova
University of Calgary, Department of Computer Science
2500 University Drive N.W., Calgary, AB, T2N 1N4, Canada
E-mail: mgavrilo@ucalgary.ca

C.J. Kenneth Tan
Exascala Ltd.
Unit 9, 97 Rickman Drive, Birmingham B15 2AL, UK
E-mail: cjtan@exascala.com

Guest Editor

François Anton
Technical University of Denmark
Department of Informatics and Mathematical Modeling
Richard Petersens Plads, 2800 Lyngby, Denmark
E-mail: fa@imm.dtu.dk

Library of Congress Control Number: 2010934647

CR Subject Classification (1998): I.3.5, G.2, J.3, F.2, F.2.2, F.4.1

ISSN	0302-9743 (Lecture Notes in Computer Science)
ISSN	1866-4733 (Transactions on Computational Science)
ISBN-10	3-642-16006-9 Springer Berlin Heidelberg New York
ISBN-13	978-3-642-16006-6 Springer Berlin Heidelberg New York

springer.com

© Springer-Verlag Berlin Heidelberg 2010
Printed in Germany

Typesetting: Camera-ready by author, data conversion by Scientific Publishing Services, Chennai, India
Printed on acid-free paper   06/3180

LNCS Transactions on Computational Science

Computational science, an emerging and increasingly vital field, is now widely recognized as an integral part of scientific and technical investigations, affecting researchers and practitioners in areas ranging from aerospace and automotive research to biochemistry, electronics, geosciences, mathematics, and physics. Computer systems research and the exploitation of applied research naturally complement each other. The increased complexity of many challenges in computational science demands the use of supercomputing, parallel processing, sophisticated algorithms, and advanced system software and architecture. It is therefore invaluable to have input by systems research experts in applied computational science research.

Transactions on Computational Science focuses on original high-quality research in the realm of computational science in parallel and distributed environments, also encompassing the underlying theoretical foundations and the applications of large-scale computation. The journal offers practitioners and researchers the opportunity to share computational techniques and solutions in this area, to identify new issues, and to shape future directions for research, and it enables industrial users to apply leading-edge, large-scale, high-performance computational methods.

In addition to addressing various research and application issues, the journal aims to present material that is validated – crucial to the application and advancement of the research conducted in academic and industrial settings. In this spirit, the journal focuses on publications that present results and computational techniques that are verifiable.

Scope

The scope of the journal includes, but is not limited to, the following computational methods and applications:

- Aeronautics and Aerospace
- Astrophysics
- Bioinformatics
- Climate and Weather Modeling
- Communication and Data Networks
- Compilers and Operating Systems
- Computer Graphics
- Computational Biology
- Computational Chemistry
- Computational Finance and Econometrics
- Computational Fluid Dynamics
- Computational Geometry

- Computational Number Theory
- Computational Physics
- Data Storage and Information Retrieval
- Data Mining and Data Warehousing
- Grid Computing
- Hardware/Software Co-design
- High-Energy Physics
- High-Performance Computing
- Numerical and Scientific Computing
- Parallel and Distributed Computing
- Reconfigurable Hardware
- Scientific Visualization
- Supercomputing
- System-on-Chip Design and Engineering

Editorial

The Transactions on Computational Science journal is part of the Springer series *Lecture Notes in Computer Science*, and is devoted to the gamut of computational science issues, from theoretical aspects to application-dependent studies and the validation of emerging technologies.

The journal focuses on original high-quality research in the realm of computational science in parallel and distributed environments, encompassing the facilitating theoretical foundations and the applications of large-scale computations and massive data processing. Practitioners and researchers share computational techniques and solutions in the area, identify new issues, and shape future directions for research, as well as enable industrial users to apply the techniques presented.

The current volume is devoted to *Voronoi diagrams in science and engineering,* edited by François Anton. It is comprised of extended versions of nine papers, carefully selected from the list of accepted and presented papers at the International Symposium on Voronoi Diagrams 2009, Copenhagen, Denmark, June 23–26, 2009.

This special issue is devoted to state-of-the-art research and a broad spectrum of applications of one of the most versatile data structures in computational geometry – the Voronoi diagram.

We would like to extend our sincere appreciation to Special Issue Guest Editor François Anton for his dedication to the preparation of this high-quality special issue, to all authors for submitting their papers, and to all Associate Editors and referees for their valuable work. We would also like to express our gratitude to the LNCS editorial staff of Springer, in particular Alfred Hofmann, Ursula Barth and Anna Kramer, who supported us at every stage of the project.

It is our hope that the fine collection of papers presented in this special issue will be a valuable resource for Transactions on Computational Science readers and will stimulate further research into the vibrant area of computational science applications.

June 2010

Marina L. Gavrilova
C.J. Kenneth Tan

Voronoi Diagrams
in Science and Engineering
Special Issue Guest Editor's Preface

Stemming from the work of the German mathematician Johann Peter Gustav Lejeune Dirichlet on quadratic forms in 2D and 3D (1850), Voronoi diagrams were further generalised to n-dimensional space by the Russian mathematician Georgy Feodosevich Voronoy (1908). Voronoi diagrams have been named after the latter author in computer science, while they were named Dirichlet tessellations in mathematics. Voronoi tesselations are decompositions of a metric space induced by a discrete set of generators (which are subsets of this metric space) into proximal regions defined by a distance. Each generator defines a Voronoi cell, which is the locus of all the points of the metric space that are closer to that generator than any other generator. The boundaries of the Voronoi cells define a diagram called the "Voronoi diagram". Voronoi diagrams have been rediscovered many times in different disciplines, giving this fundamental data structure different names in each one of these disciplines: Thiessen polygons in geophysics and meteorology and Wigner-Seitz cells in the study of crystallographic structures in solid-state physics. Voronoi diagrams have been used in a wide variety of fields ranging from astronomy (Descartes's diagram of the structure of the universe in 1644), bioinformatics, computational chemistry, computer graphics, geography, material science, medicine (the British physician John Snow's illustration of the infected pump in the Soho cholera epidemic) to robotics. Voronoi diagrams have been used to explain phenomena that are related to proximity. Franz Aurenhammer has called Voronoi diagrams "the universal spatial data structure". The dual graph of the Voronoi diagram is called a Delaunay triangulation after Boris Nikolaevich Delaunay (or Delone), who was a PhD student of Georgy Feodosevich Voronoy at Kiev University. It stores the topology of the Voronoi tessellation.

This special issue on Voronoi Diagrams in Science and Engineering in Springer's Transactions on Computational Science (TCS) journal presents a snapshot of current theoretical and applied research on Voronoi diagrams. It is devoted to the 6th annual International Symposium on Voronoi Diagrams in Science and Engineering (ISVD), held at the Technical University of Denmark, Kongens Lyngby, Lyngby-Taarbæk municipality, Denmark, June 23–26, 2009, and chaired by François Anton.

The ISVD symposium was initially created due to the popular demand among the scientific community for an international forum devoted mainly to Voronoi diagrams and Delaunay triangulations. ISVD was initiated by Kokichi Sugihara at the University of Tokyo, Japan in 2004. Since then, ISVD has been held every year: Deok-Soo Kim chaired ISVD 2005 in Seoul, Korea; Marina Gavrilova chaired ISVD 2006 in Banff, Alberta, Canada; Christopher Gold chaired ISVD 2007 in Pontypridd, Wales, UK; Victor P. Andrushchenko, Anatoly M. Samoilenko, and Valerii V. Kozlov chaired ISVD 2008 in Kiev, Ukraine jointly with the Fourth International Kyiv Conference on Analytical Number Theory and Space Tessellations.

Nine papers were selected from ISVD 2009, extended to journal papers, reviewed after extension and re-reviewed if needed after corrections. These papers, published in the present special issue, cover different aspects of Voronoi diagrams, which can be grouped in the following way:

- divide and conquer construction of Voronoi diagrams (from envelopes in space);
- new generalised Voronoi diagrams or properties of existing generalised Voronoi diagrams (reversibility in the Voronoi data structure of points and line segments, round-tour Voronoi diagrams, triangle-perimeter two-site Voronoi diagrams);
- applications of Voronoi diagrams and their duals in graph theory (approximating shortest path queries), computer graphics (visual hull reconstruction), bioinformatics (protein-ligand docking) and spatial process simulation.

The editor expects that the readers of Transactions on Computational Science (TCS) will benefit from reading the papers presented in this special issue on the latest advances in Voronoi diagrams in science and engineering.

The guest editor of the special issue on Voronoi Diagrams in Science and Engineering in Springer's Transactions on Computational Science (TCS) and chair of ISVD 2009 would like to thank all invited authors for submitting an extended version of their ISVD 2009 paper and all the reviewers and members of the Programme Committee of ISVD 2009 for their great commitment in reviewing the papers of ISVD 2009 and of this special issue. This special issue would not have been possible without the active help, commitment and support of my colleagues in the Steering Committee of ISVD, the active help, commitment and support through all the stages of the special issue of the Editor-in-Chief, Marina L. Gavrilova, the vision and commitment of the other Editor-in-Chief, Chih Jeng Kenneth Tan, and the professional help and commitment from the publisher Springer.

June 2010   François Anton
                                                                      Guest Editor

LNCS Transactions on
Computational Science –
Editorial Board

Table of Contents

Constructing Two-Dimensional Voronoi Diagrams via
Divide-and-Conquer of Envelopes in Space 1
    Ophir Setter, Micha Sharir, and Dan Halperin

Approximate Shortest Path Queries Using Voronoi Duals 28
    *Shinichi Honiden, Michael E. Houle, Christian Sommer, and
    Martin Wolff*

On the Triangle-Perimeter Two-Site Voronoi Diagram 54
    Iddo Hanniel and Gill Barequet

Voronoi Graph Matching for Robot Localization and Mapping 76
    Jan Oliver Wallgrün

Properties and an Approximation Algorithm of Round-Tour Voronoi
Diagrams 109
    Hidenori Fujii and Kokichi Sugihara

Protein-Ligand Docking Based on Beta-Shape 123
    *Chong-Min Kim, Chung-In Won, Jae-Kwan Kim, Joonghyun Ryu,
    Jong Bhak, and Deok-Soo Kim*

Kinetic Line Voronoi Operations and Their Reversibility 139
    Darka Mioc, François Anton, Christopher Gold, and Bernard Moulin

High Quality Visual Hull Reconstruction by Delaunay Refinement...... 166
    Xin Liu and Marina L. Gavrilova

Geosimulation of Geographic Dynamics Based on Voronoi Diagram..... 183
    Mir Abolfazl Mostafavi, Leila Hashemi Beni, and Karine Hins Mallet

Author Index 203

Constructing Two-Dimensional Voronoi Diagrams via Divide-and-Conquer of Envelopes in Space*

Ophir Setter, Micha Sharir, and Dan Halperin

School of Computer Science
Tel-Aviv University
Tel-Aviv, Israel
{ophirset,michas,danha}@post.tau.ac.il

Abstract. We present a general framework for computing Voronoi diagrams of different classes of sites under various distance functions in $\mathbb{R}^2$. Most diagrams mentioned in the paper are in the plane. However, the framework is sufficiently general to support diagrams embedded on a family of two-dimensional parametric surfaces in three-dimensions. The computation of the diagrams is carried out through the construction of envelopes of surfaces in 3-space provided by CGAL (the Computational Geometry Algorithm Library). The construction of the envelopes follows a divide-and-conquer approach. A straightforward application of the divide-and-conquer approach for Voronoi diagrams yields algorithms that are inefficient in the worst case. We prove that through randomization, the expected running time becomes near-optimal in the worst case. We also show how to apply the new framework and other existing tools from CGAL to compute minimum-width annuli of sets of disks, which requires the computation of two Voronoi diagrams of different types, and of the overlay of the two diagrams. We do not assume general position. Namely, we handle degenerate input, and produce exact results.

1 Introduction

Voronoi diagrams were thoroughly investigated, and were used to solve many geometric problems, since introduced by Shamos and Hoey to the field of

---

* Work on this paper has been supported in part by the Hermann Minkowski–Minerva Center for Geometry at Tel-Aviv University. Work by Ophir Setter and Dan Halperin has also been supported in part by the Israel Science Foundation (Grant no. 236/06), and by the German-Israeli Foundation (Grant no. 969/07). Work by Micha Sharir was also partially supported by NSF Grants CCF-05-14079 and CCF-08-30272, by Grant 2006/194 from the U.S.-Israeli Binational Science Foundation, by Grants 155/05 and 338/09 from the Israel Science Fund, Israeli Academy of Sciences, and by a grant from the French-Israeli AFIRST program.

M.L. Gavrilova et al. (Eds.): Trans. on Comput. Sci. IX, LNCS 6290, pp. 1–27, 2010.
© Springer-Verlag Berlin Heidelberg 2010

computer science [1] (although their origin dates back centuries ago — see [2]). The concept of Voronoi diagrams was extended to handle various kinds of geometric sites, ambient spaces, and distance functions. Among those are power diagrams of points in the plane, multiplicatively-weighted Voronoi diagrams, additively-weighted Voronoi diagrams (also known as Apollonius diagrams), Voronoi diagrams of line segments, and many other types [2, 3, 4]. There are types of Voronoi diagrams that are defined for pairs (or larger tuples) of sites [5]. Different types of Voronoi diagrams have been unified under a generalized framework [6].

Numerous approaches for computing Voronoi diagrams were developed: the divide-and-conquer algorithm by Shamos and Hoey [1], the sweep-line algorithm by Fortune [7], randomized incremental constructions [8, 9], a dynamic algorithm for planar convex objects [10], and more. A recent approach employs a divide-and-conquer medial-axis algorithm on an augmented domain to compute the Euclidean Voronoi diagrams of various types of sites [11]. Available exact implementations of Voronoi diagram algorithms include the computation of Delaunay graphs dual to standard Voronoi diagrams, to Apollonius diagrams, and to segment Voronoi diagrams in CGAL [12, 13, 14], the Voronoi diagrams of ellipses in CGAL [15], and the construction of segment Voronoi diagrams in LEDA [16]. Prominent approximated alternatives include the VRONI code that uses floating-point arithmetic to compute Voronoi diagrams of points, segments, and arcs in 2D [17], and other implementations that use the Graphics Processing Unit (GPU) to discretely compute Voronoi diagrams [18, 19]. Typically, the time complexity of constructing a Voronoi diagram that has linear complexity, using the above algorithms, is nearly linear.

Edelsbrunner and Seidel observed the connection between Voronoi diagrams in $\mathbb{R}^d$ and lower envelopes of the distance functions that correspond to the sites in $\mathbb{R}^{d+1}$ (see also Section 2), yielding a very general approach for computing Voronoi diagrams [20].

The Computational Geometry Algorithms Library (CGAL)[1] contains a robust and efficient implementation of a divide-and-conquer algorithm for constructing envelopes of surfaces in three dimensions [21, 22]. The theoretical worst-case time complexity of constructing the envelope of n "well-behaved" surfaces in three dimensions using the algorithm is[2] $O(n^{2+\varepsilon})$ [23] (this is also an upper bound, almost tight in the worst case, on the combinatorial complexity of the envelope). As observed below, this near-quadratic running time can arise also in the case of envelopes that represent Voronoi diagrams of linear complexity. This fact poses an obstacle when this divide-and-conquer algorithm is used for computing Voronoi diagrams that have linear complexity, as we aim for algorithms that run in near-linear time. We show, in Section 3, that using randomization in the

---

[1] http://www.cgal.org

[2] A bound of the form $O(f(n) \cdot n^\varepsilon)$ means that the actual upper bound is $C_\varepsilon f(n) \cdot n^\varepsilon$, for any $\varepsilon > 0$, where C_ε is a constant that depends on ε, and generally tends to infinity as ε goes to 0.

divide step eliminates the high-complexity cost of this approach, and yields an algorithm with asymptotically efficient expected running time.

We present a general framework for computing various two-dimensional Voronoi diagrams, exploiting the efficient, robust, and general-purpose code of CGAL for constructing envelopes of surfaces. Section 4 explains at a high level how to produce new types of Voronoi diagrams, and shows various examples of instantiations of our framework for computing numerous types of Voronoi diagrams, together with other experimental results. Appendix A supplies technical details on the software interface and outlines the concrete steps for producing code for new types of Voronoi diagrams. Further details on the implementation can be found in [24]. The framework is sufficiently general to support diagrams embedded on certain two-dimensional orientable parametric surfaces in $\mathbb{R}^3$, exploiting the facts that the diagrams can be represented as arrangements on the given surface, and that CGAL contains a package that facilitates operations on such arrangements [25]. Section 5 generalizes an algorithm for computing a minimum-width annulus of a set of points in the plane [26] to a set of disks in the plane, and shows how the new framework for Voronoi diagrams can be applied to successfully solve this problem.

The major strength of our approach is its completeness, robustness, and generality, that is, the ability to handle degenerate input, the agility to produce exact results, and the capability to construct diverse types of Voronoi diagrams. The code is designed to successfully handle degenerate input, while exploiting the synergy between generic programming and exact geometric computing, and the divide-and-conquer framework to construct Voronoi diagrams. From a theoretical point of view, the randomized divide-and-conquer envelope approach for computing Voronoi diagrams is shown to be efficient (in an expected sense) and to be asymptotically comparable to other (near-)optimal methods. However, the method uses constructions of bisectors and Voronoi vertices as elementary building blocks, and they must be exact, which makes the concrete running time of our exact implementation inferior to various existing implementations dedicated to specific diagram-types that do not necessarily construct bisectors and vertices (e.g., computing the Delaunay graphs dual to Voronoi diagrams does not require construction of bisectors).

Our software practically supports any kind of nearest-site Voronoi diagrams as well as the corresponding farthest-site Voronoi diagrams, provided that the user supplies a set of basic procedures for manipulating a small number of sites and their bisectors. The implementation of these procedures is an art in itself. However, recent tools for manipulating arbitrary algebraic plane-curves [27, 28] cover a wide range of bisector types, which enables the construction of different types of Voronoi diagrams with greater ease. Table 1 lists the types of diagrams that are currently supported by our implementation. Figure 1 illustrates several types of planar Voronoi diagrams computed with our software. Figure 2 shows two types of Voronoi diagrams on the sphere computed with our software. Both diagrams are composed of geodesic arcs.

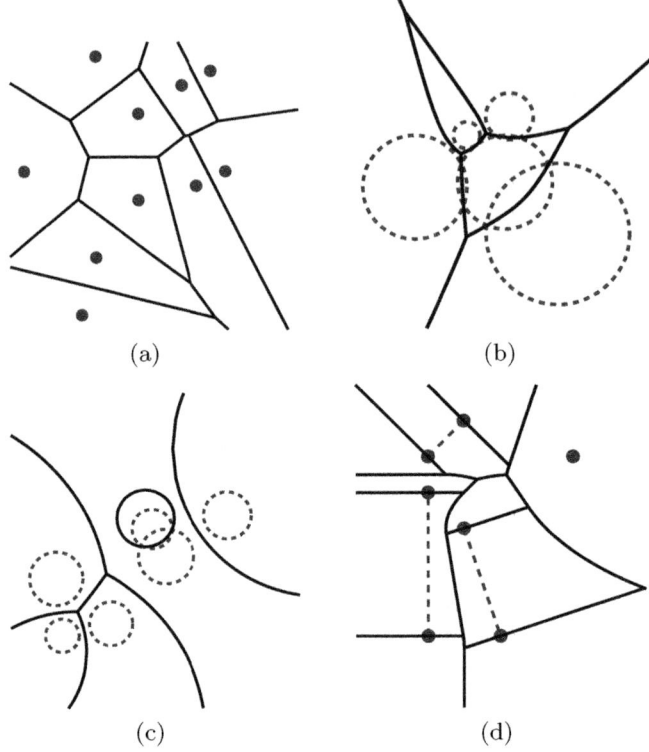

(a)   (b)

(c)   (d)

Fig. 1. Various Voronoi diagrams computed with our software. (For the parameters of sites in each diagram, see Table 1.) (a) A standard Voronoi diagram (point sites with L_2 metric). (b) An Apollonius diagram (additively-weighted Voronoi diagram) with disk centers as sites and disk radii as weights. (c) A Möbius diagram with disk centers as sites. The distance from every point on the boundary of a disk to its corresponding site is zero. (d) A Voronoi diagram of segments and points. The sites in (b), (c), and (d) are illustrated with dashed curves.

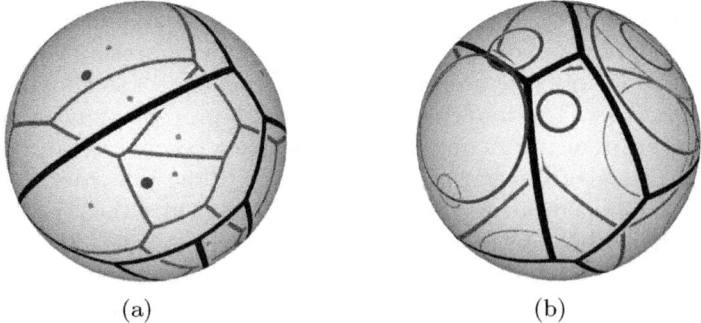

(a)   (b)

Fig. 2. Voronoi diagrams on the sphere computed with our software. (a) Spherical Voronoi diagram of 14 points. (b) Spherical power diagram of 10 circular sites.

Table 1. Types of Voronoi diagrams currently supported by our implementation, and their bisector classes

Name	Sites	Distance function	Bisector class
Standard Voronoi diagram	points p_i	$\|x - p_i\|$	lines
Power diagram	disks (with center c_i and radius r_i)	$\sqrt{(x - c_i)^2 - r_i^2}$	
2-point triangle-area Voronoi diagram	pairs of points $\{p_i, q_i\}$	area of $\triangle x p_i q_i$	pairs of lines
Apollonius diagram	points p_i and weights w_i	$\|x - p_i\| - w_i$	hyperbolic arcs
Möbius diagram	points p_i with scalars λ_i, μ_i	$\lambda_i (x - p_i)^2 - \mu_i$	circles and lines
Anisotropic diagram	points p_i, with positive definite matrices M_i, and a scalar π_i	$(x - p_i)^t M_i (x - p_i) - \pi_i$	conic arcs
Voronoi diagram of linear objects	interior-disjoint points, segments, rays, or lines	Euclidean distance	piecewise algebraic curves composed of line segments and parabolic arcs
Spherical Voronoi diagram	points on a sphere	geodesic distance	arcs of great circles (geodesic arcs)
Power diagram on a sphere	circles on a sphere	"spherical" power distance[a]	

[a] Given a point p and a circle with center q and radius r on the sphere, the spherical power "proximity" between p and the circle is defined to be $\frac{\cos d(p,q)}{\cos r}$ where $d(p,q)$ is the geodesic distance between p and q [29].

2   Preliminaries

2.1   A Divide-and-Conquer Algorithm for Constructing Voronoi Diagrams

Definition 1 (Lower envelope). *Given a set of bivariate functions $F = \{f_1, \ldots, f_n\}$, $f_i : \mathbb{R}^2 \rightarrow \mathbb{R}$ for each i, their* lower envelope *$\Psi(x,y)$ is defined to be their pointwise minimum:*

$$\Psi(x,y) = \min_{1 \leq i \leq n} f_i(x,y).$$

The *minimization diagram* of F is the subdivision of the xy-plane into maximal relatively-open connected cells, such that the function (or the set of functions) that attains the lower envelope over a specific cell of the subdivision is the same for all points in the cell.

Definition 2 (Voronoi diagram). *Let $O = \{o_1, \ldots, o_n\}$ be a set of n objects in the plane (also called* Voronoi sites*). The* Voronoi diagram $\mathrm{Vor}_\rho(O)$ *of O with respect to a given distance function ρ, is the partition of the plane into maximal* *relatively-open connected cells, where each cell consists of points that are closer to one particular site (or a set of sites) than to any other site.*

In certain cases, the distance to a site may depend on various parameters associated with the site; see, for example, the cases of Möbius diagrams or anisotropic diagrams [4] in Table 1. The *bisector* $B(o_i, o_j)$ of two Voronoi sites $o_i, o_j \in O$ is the locus of all points that have an equal distance to both sites, that is

$$B(o_i, o_j) = \{x \in \mathbb{R}^2 \mid \rho(x, o_i) = \rho(x, o_j)\}.$$

From the above definitions it is clear that if we let $f_i : \mathbb{R}^2 \to \mathbb{R}$ to be $f_i(x) = \rho(x, o_i)$, for each $i = 1, \dots, n$, then the minimization diagram of $\{f_1, \dots, f_n\}$ is exactly the Voronoi diagram of O.

Constructing envelopes of bivariate functions in $\mathbb{R}^3$ can be done with a divide-and-conquer algorithm [23]. The algorithm is adapted to Voronoi diagrams computations as follows:[3] let S be a collection of n sites in the plane and let ρ be some distance function. We partition S into two disjoint subsets S_1 and S_2 of (roughly) equal size, recursively construct their respective Voronoi diagrams $\text{Vor}_\rho(S_1)$ and $\text{Vor}_\rho(S_2)$, and then merge the two diagrams to obtain $\text{Vor}_\rho(S)$.

The merging step starts with overlaying $\text{Vor}_\rho(S_1)$ and $\text{Vor}_\rho(S_2)$. For each face f of the overlay, all its points have a fixed pair of nearest sites s_1 and s_2 from S_1 and S_2, respectively, and the bisector between s_1 and s_2 (restricted to f) partitions f into its portion of points nearer to s_1 and the complementary portion of points nearer to s_2. This results with pieces of the final Voronoi cells. Each feature of the refined overlay is labeled with the site nearest to it. Finally, redundant features are removed (these are vertices and portions of edges from one Voronoi diagram that lie closer to a site in the other Voronoi diagram), and subcells of the same cell are stitched together, to yield the combined final diagram. Figure 3 illustrates the merging step of two Voronoi diagrams.

The asymptotic worst-case time complexity of the divide-and-conquer envelope algorithm (under the natural assumption that the objects and distance function have "constant description complexity") is $O(n^{2+\varepsilon})$, for any $\varepsilon > 0$. Indeed, there are Voronoi diagrams in the plane that have quadratic complexity for which this construction is nearly worst-case optimal. An example for such diagrams are multiplicatively-weighted Voronoi diagrams [30] (see Figure 4 for an illustration). However, in cases where the complexity of the diagrams is sub-quadratic (for most of the cases, linear), we would like the algorithm to run in sub-quadratic (or near-linear) time.

2.2   Constructing Envelopes in CGAL

CGAL is a library of efficient and reliable geometric data structures and algorithms. CGAL follows the exact geometric computation paradigm [31], and

---

[3] The description assumes general position, even though the algorithm, as well as its implementation, also handle degenerate cases.

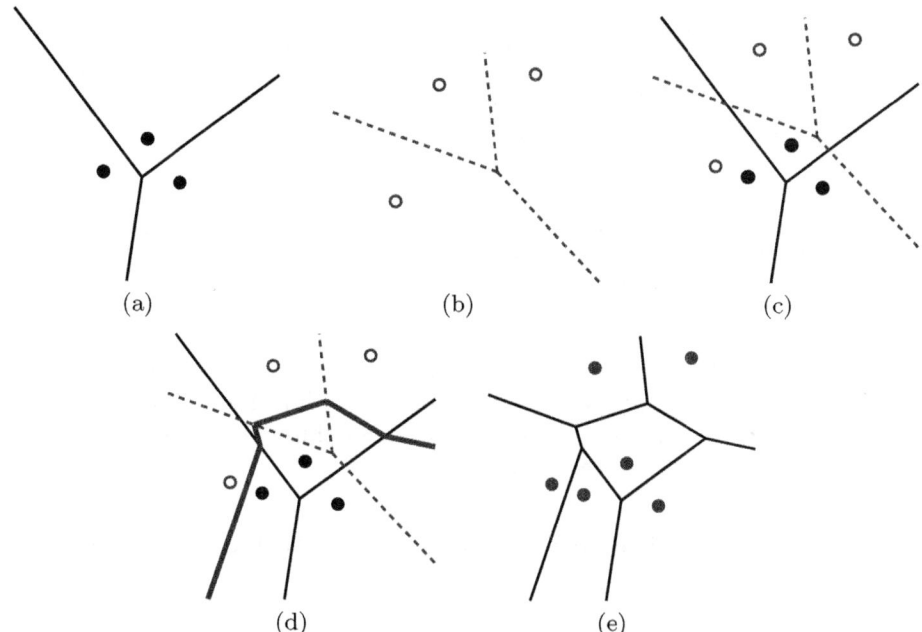

Fig. 3. The merge step of the divide-and-conquer algorithm for computing Voronoi diagrams. (a) The first Voronoi diagram, $\text{Vor}_\rho(S_1)$. (b) The second Voronoi diagram, $\text{Vor}_\rho(S_2)$. (c) The overlay of the two diagrams. (d) The refined overlay. Each face is partitioned to regions that are closer to sites from S_1 and region that are closer to sites from S_2. (e) The final diagram obtained after the removal of redundant features from the refined overlay, and the stitching of the remaining pieces.

adheres to the generic programming paradigm [32] to achieve maximum flexibility without compromising efficiency.

The Arrangement_on_surface_2 package of CGAL supports the construction, the maintenance, and the manipulation of arrangements embedded on certain two-dimensional oriented parametric surfaces in three dimensions, such as spheres, cylinders, tori, etc. [25, 33].

The arrangement package is the basis of the Envelope_3 package of CGAL, which implements the algorithm mentioned in the previous section for computing the lower (or the upper) envelope of a set of surfaces in three dimensions [21]. While insuring stability, the number of calls to the exact (and slow) geometric predicates is minimized by propagating pre-computed information about the structure of the envelope to neighboring cells in the merge step of the algorithm. The package decouples the topology-related computation from the geometry-related computation, resulting in a generic software which is easy to reuse and adapt. The Envelope_3 package handles all degenerate situations, and when used with exact numeric types, achieves robustness and produces exact results.

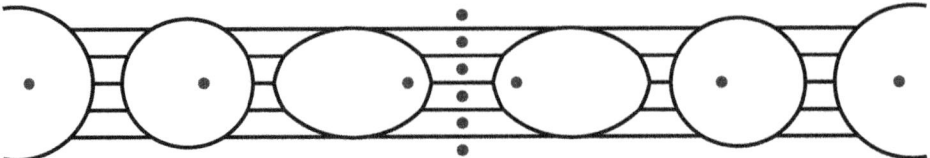

Fig. 4. A worst-case quadratic-size example of a multiplicatively-weighted Voronoi diagram with 12 sites, based on an example by Aurenhammer and Edelsbrunner [30]. The diagram was computed with our software.

The aforementioned definitions (Section 2.1) can be generalized to lower envelopes (or Voronoi diagrams) projected onto two-dimensional parametric surfaces. Furthermore, the divide-and-conquer algorithm can be used to compute lower (or upper) envelopes over these surfaces.

3   Near-Optimal Bound on the Expected Construction Time

The complexity of the merge step of the algorithm described in Section 2.1 directly depends on the complexity of the overlay of the two partial diagrams. (The cost of the best general algorithm for constructing the overlay is larger by a logarithmic factor than the combined complexity of the input diagrams and of the overlay.) Careless partition of the input sites into two subsets can dramatically slow down the computation. For example, consider the standard L_2-diagram for the following point-set in the plane, $S = \{(i, i)\}_{i=1}^{n/2} \cup \{(-i, i)\}_{i=1}^{n/2}$; see Figure 5(a) for an illustration. If we partition the set into two subsets, to the left and to the right of the y-axis ("deterministic" partitioning strategy), then in the final merge step, the overlay of the two sub-diagrams has $\Theta(n^2)$ complexity. Hence the algorithm runs in $\Omega(n^2)$ time, even though the complexity of the final diagram is only $\Theta(n)$. We argue that when the partitioning is done *randomly*, the expected complexity of the overlay is comparable with the maximum complexity of the diagram for essentially any kind of sites and distance function, and for any possible input. The proof assumes general position, but the results remain true for degenerate cases as well.

Theorem 1. *Consider a specific type of two-dimensional Voronoi diagrams, so that the worst-case complexity of the diagram of any set of at most n sites is $F(n)$. Let S be a set of n sites. If we randomly split S into two subsets S_1 and S_2, by choosing at random for each site, with equal probability, the subset it belongs to, then the expected complexity of the overlay of the Voronoi diagram of S_1 with the Voronoi diagram of S_2 is $O(F(n))$.*

Proof. Each vertex of the overlay is either a vertex of $\mathrm{Vor}_\rho(S_1)$, a vertex of $\mathrm{Vor}_\rho(S_2)$, or a crossing between an edge of $\mathrm{Vor}_\rho(S_1)$ and an edge of $\mathrm{Vor}_\rho(S_2)$ (in degenerate cases, this is a non exclusive disjunction). The number of vertices of the first two kinds is $O(F(n))$, so it suffices to bound the expected number of crossings between edges of $\text{Vor}_\rho(S_1)$ and of $\text{Vor}_\rho(S_2)$. Such a crossing is a point u that is defined by four sites, $p_1, q_1 \in S_1$ and $p_2, q_2 \in S_2$, so that u lies on the Voronoi edge $e(p_1, q_1)$ of $\text{Vor}_\rho(S_1)$ that bounds the cells of p_1 and q_1, and on the Voronoi edge $e(p_2, q_2)$ of $\text{Vor}_\rho(S_2)$ that bounds the cells of p_2 and q_2. Without loss of generality, assume that $\rho(u, p_1) = \rho(u, q_1) < \rho(u, p_2) = \rho(u, q_2)$.

A simple but crucial observation is that u must also lie on the Voronoi edge between the cells of p_1, q_1 in the *overall* diagram $\text{Vor}_\rho(S)$. Indeed, if this were not the case then there must exist another site $s \in S$ so that u is nearer to s than to p_1, q_1. But then s cannot belong to S_1, for otherwise it would prevent u from lying on the Voronoi edge of p_1, q_1. For exactly the same reason, s cannot belong to S_2 — it would then prevent u from lying on the Voronoi edge of p_2, q_2 in that diagram. This contradiction establishes the claim.

Define the *weight* k_u of u to be the number of sites s satisfying

$$\rho(u, p_1) = \rho(u, q_1) < \rho(u, s) < \rho(u, p_2) = \rho(u, q_2).$$

Clearly, all these k_u sites must be assigned to S_1.

In other words, for any crossing point u between two Voronoi edges $e(p_1, q_1)$, $e(p_2, q_2)$, with weight k_u (with all the corresponding k_u sites being farther from u than p_1, q_1 and nearer than p_2, q_2), u appears as a crossing point in the overlay of $\text{Vor}_\rho(S_1)$, $\text{Vor}_\rho(S_2)$ if and only if the following three conditions (or their symmetric counterparts, obtained by reversing the roles of S_1 and S_2) hold: (i) $p_1, q_1 \in S_1$; (ii) $p_2, q_2 \in S_2$; and (iii) all the k_u sites that contribute to the weight are assigned to S_1. This happens with probability $\frac{1}{2^{k_u + 3}}$.

Hence, if we denote by N_w (resp., $N_{\leq w}$) the number of crossings of weight w (resp., of weight at most w), the expected number of crossings in the overlay is

$$\sum_{w \geq 0} \frac{N_w}{2^{w+3}} = O\left(\sum_{w \geq 0} \frac{N_{\leq w}}{2^w}\right), \tag{1}$$

where the right-hand side is obtained by substituting $N_w = N_{\leq w} - N_{\leq w-1}$, and by a simple rearrangement of the sum.

We can obtain an upper bound on $N_{\leq w}$ using the Clarkson-Shor technique [34]. Specifically, denote by $N_w(n)$ (resp., $N_{\leq w}(n)$) the maximum value of N_w (resp., $N_{\leq w}$), taken over all sets of n sites in the class under consideration. Then, since a crossing is defined by four sites, we have

$$N_{\leq w}(n) = O\left(w^4 N_0(n/w)\right).$$

Note that if a crossing u, defined by p_1, q_1, p_2, q_2, has weight 0 then p_1, q_1, p_2, q_2 are the four nearest sites to u. The number of such quadruples is thus upper bounded by the complexity of the *fourth-order* Voronoi diagram of some set (actually, a random sample) S_0 of n/w sites.

We claim that the complexity of the fourth-order Voronoi diagram of n sites is $O(F(n))$. Indeed, any quadruple p_1, q_1, p_2, q_2 of four nearest sites to some point u can be charged (essentially, in a unique manner) to a face of the fourth-order diagram (the one containing u). Each such face can in turn be charged either to one of its vertices, or to its rightmost point, or to a point at infinity on one of its edges. Assuming general position, each such boundary point can be charged at most $O(1)$ times. Now, another simple application of the Clarkson-Shor technique shows that the number of these vertices and boundary points is $O(F(n))$ — each of them becomes a feature of the (0-order) Voronoi diagram if we remove a constant number of sites, which happens with large probability when we sample a constant fraction of the sites.

In other words, we have $N_{\leq w}(n) = O(w^4 F(n/w)) = O(w^4 F(n))$. Substituting this into (1), we obtain an upper bound of $O(F(n))$ on the complexity of the overlay, as claimed.   □

As we aim to compute Voronoi diagrams of a large variety of types, we use a sweepline based overlay algorithm that exhibits good practical performance, and incurs a mere logarithmic factor over the optimal computing time, namely $O(F(n) \log n)$. In particular we use the overlay operation provided by the `Arrangement_on_surface_2` package. Combined with the "master recurrence" $T(n) = 2T(\frac{n}{2}) + O(F(n) \log n)$ for the expected running time $T(n)$ of the algorithm, we obtain the following special cases:

Corollary 1. *For any specific type of two-dimensional Voronoi diagrams, so that the worst-case complexity of the diagram of any set of at most n sites is $O(n)$, the divide-and-conquer envelope algorithm computes the diagram in expected $O(n \log^2 n)$ time. If the worst-case complexity $F(n)$ is $\Omega(n^{1+\varepsilon})$ then the expected running time is $O(F(n) \log n)$.*

When the diagram is a convex subdivision, one can carry out the merge step more efficiently, in linear $O(F(n))$ expected time using the procedure described by Guibas and Seidel [35]. In particular, we have:

Corollary 2. *The L_2-Voronoi diagram of n points in the plane, or the power diagram of n disks in the plane, can be computed using the randomized divide-and-conquer envelope algorithm in expected optimal $O(n \log n)$ time.*

Remark 1. The analysis used to prove Theorem 1 can easily be extended to the case of lower envelopes of arbitrary collections of bivariate functions (of constant description complexity). As a result, we get the following:

Corollary 3. *Let $\mathcal{G}$ be a collection of n bivariate functions of constant description complexity each, and let $F(m)$ be an upper bound on the complexity of the lower envelope of any subcollection of at most m functions. Then the expected complexity of the overlay of the minimization diagrams of two subcollections $\mathcal{G}_1$ and $\mathcal{G}_2$, obtained by randomly partitioning $\mathcal{G}$, as above, is $O(F(n))$. Consequently, the lower envelope of $\mathcal{G}$ can be constructed by the above randomized divide-and-conquer technique, in expected time $O(F(n) \log n)$, provided that*

$F(n) = \Omega(n^{1+\varepsilon})$, *for some $\varepsilon > 0$. The expected running time is $O(n \log^2 n)$ when* $F(n) = O(n)$.

4   Examples and Experimental Results

Figure 5 demonstrates the effect of executing the recursion by randomly partitioning the sites into two subsets of equal size, when constructing standard Voronoi diagrams with the worst-case inputs mentioned in Section 3. If we partition the set into two subsets, to the left and to the right of the y-axis ("deterministic" partitioning strategy), the overlay of the two Voronoi diagrams has $\Theta(n^2)$ complexity. Figure 5(b) illustrates the fact, established above, that partitioning the sites randomly, results in running time nearly-linear in the number of sites.

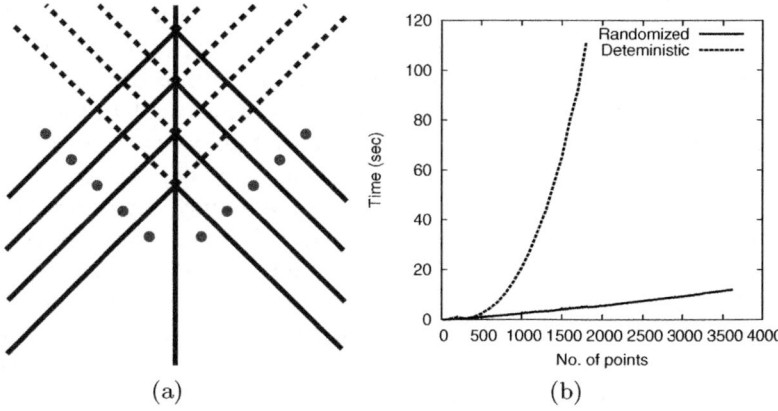

(a)   (b)

Fig. 5. Effect of randomization. (a) A worst-case example of 10 point sites in the case of the standard Voronoi diagram. The dashed segments are segments that exist in the overlay (if we use the deterministic partition strategy) but will not be present in the final diagram. (b) The graph shows the running time in seconds as a function of the number of sites (arranged in a worst case constellation as in (a)) in both the worst-case deterministic and the randomized partitioning strategies.

A prime advantage of our framework is the ability to compute new types of Voronoi diagrams with relative ease. Despite the fact that the envelope code is used to compute Voronoi diagrams, the user does not have to define the surfaces explicitly, and does not have to know the algorithmic details of constructing envelopes of surfaces and their minimization diagrams. Creating a new type of Voronoi diagrams amounts only to the provision of a small set of geometric types and operations on a small number of sites and bisectors (e.g., determine which is the closer site, among two given sites, to a given point, construct the bisector of two sites, etc.), gathered in a traits class; see Appendix A for further details.

The traits class of a new type of diagrams is based on a traits class for the arrangement package, which supplies the handling and manipulation of bisector curves of pairs of sites. A Voronoi diagram, the bisectors of which can be represented with an existing traits class for the arrangement package can be easily developed. Existing traits classes support bounded or unbounded linear curves (line segments, rays, and lines), circles, circular and linear segments, conic arcs [33], Bézier curves, and real algebraic plane curves of arbitrary degree [27, 28]. These traits classes enable the construction of all the Voronoi diagrams shown in Figure 1 and listed in Table 1, and of further types of Voronoi diagrams with bisectors that can be represented by curves supported by these traits classes.

The framework supports a vast variety of Voronoi diagrams with different properties. Existing frameworks for computing Voronoi diagrams usually support a specific family of Voronoi diagrams, e.g., planar Voronoi diagrams of points under the L_p metric, planar Voronoi diagrams of general sites under the Euclidean metric, etc. Using an envelope-construction based algorithm allows our framework to support many families of two-dimensional Voronoi diagrams; linear Voronoi diagrams as well as Voronoi diagrams with quadratic complexity, and Voronoi diagrams with two-dimensional bisectors can be implemented. The diagrams do not have to conform with the definition of abstract Voronoi diagrams [6]. For example, anisotropic diagrams, which violate the theoretical requirement from an abstract Voronoi diagram that all bisectors divide the plane into two unbounded regions, and easily constructed by our approach.

The implementation of the `Envelope_3` package mainly makes use of two operations supported by the arrangement package: (i) sweep-based overlay operation, which is used to overlay two planar minimization diagrams, and (ii) zone computation-based insertion operation, which is used to insert projected intersection curves (of the surfaces) into the current planar diagram, which refine it by partitioning some of its cells. The new `Arrangement_on_surface_2` package extends the aforementioned operations, that is, the sweep-line and zone-computation, to support arrangements on two-dimensional parametric surfaces [25]. Thus, we extensively reuse code developed for planar Voronoi diagrams to compute Voronoi diagrams embedded on the sphere by extending the `Envelope_3` code to work together with the new `Arrangement_on_surface_2` package, and handle minimization diagrams that are embedded on two-dimensional parametric surfaces. We compute Voronoi diagrams of points on a sphere and power diagrams on a sphere using the framework together with the traits class for arcs of great circles on a sphere (see [36, 37] for more information, and Figures 2 and 6(d) for illustrations).

Our framework, like the `Arrangement_on_surface_2` package of CGAL, handles all degenerate situations that arise in the computation of Voronoi diagrams, as long as the supplied traits class handles a small number of degenerate cases. For example, the traits class should be able to detect whether two curves overlap, when trying to intersect them. Figure 6 depicts various degenerate scenarios that are properly handled by our framework and the traits classes. Figure 7 shows scenarios of Voronoi diagrams of points, linear segments, and lines containing certain degeneracies.

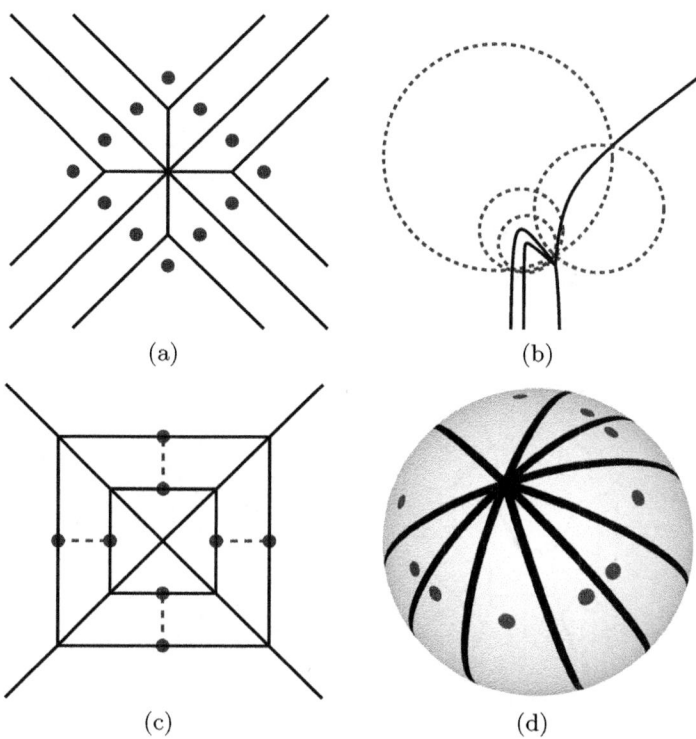

Fig. 6. Degenerate Voronoi diagrams computed with our software. (a) A degenerate standard Voronoi diagram. (b) A degenerate Apollonius (additively-weighted Voronoi) diagram. (c) A degenerate Voronoi diagram of segments and points. (d) A degenerate spherical Voronoi diagram. The sites in (b) and (c) are illustrated with dashed curves.

Given a traits class for nearest-site Voronoi diagrams, our framework can also be used to compute the respective farthest-site Voronoi diagrams by constructing upper, rather than lower, envelopes. Figure 8 shows various farthest-site Voronoi diagrams computed with our framework.

Voronoi diagrams are represented as exact CGAL arrangements, and their vertices, edges, and faces can easily be traversed while obtaining the coordinates of the vertices of the diagram to any desired precision, if the situation so requires.

The computed diagrams can be passed as input to consecutive operations supported by the `Arrangement_on_surface_2` package and its derivatives. We get plenty of additional functionalities for free. Among those are (i) point-location queries and vertical-ray shooting, (ii) the ability to perform aggregated insertion of additional curves, (iii) computing the zone of a curve inside a Voronoi diagram and inserting curves in an incremental fashion to an existing diagram, (iv) the ability to remove existing edges of the diagram, (v) overlaying two (or more) Voronoi diagrams or arrangements, (vi) the ability to attach user-defined data to all the geometric objects (i.e., points and curves) and to all the topological primitives (i.e., vertices, edges, and faces) required in certain applications, and

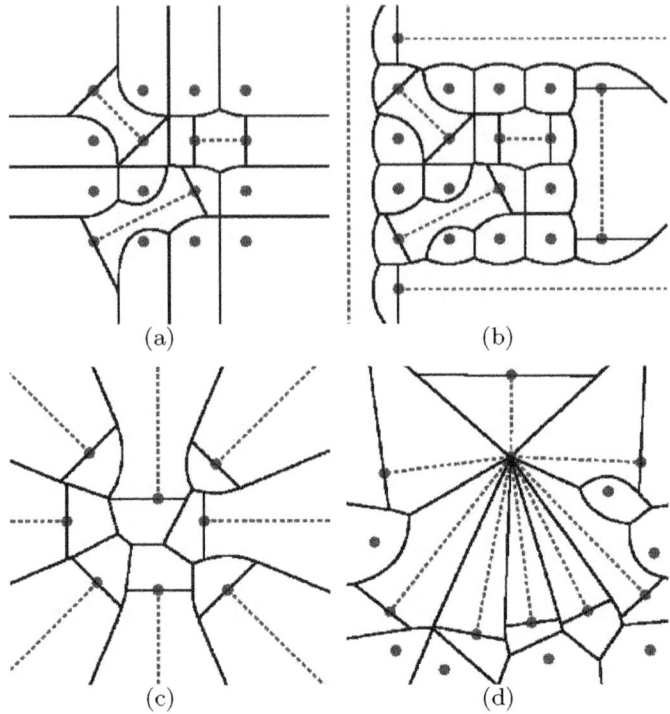

Fig. 7. Voronoi diagrams of various sets of linear objects as induced by the Euclidean metric computed with our framework. The sites are illustrated with dashed curves. (a) The Voronoi diagram of 4×4 grid with three line segments connecting 6 grid points. (b) The Voronoi diagram of a set of sites identical to the set in (a) with a line, a segment, and two rays around the grid points. (c) The Voronoi diagrams of 8 rays in the plane. (d) The Voronoi diagram of 8 segments and 7 isolated points. All segments intersect at one point, which is one of their endpoints.

(vii) the ability to use the BOOST graph library to apply various graph algorithms on the diagram and on its dual structure [33]. For example, we can use the various point-location algorithms that come with the arrangement package to solve the corresponding post-office problems [38].

Overlaying two (or more) Voronoi diagrams is immediate, a fact that we used to implement the algorithm for computing a minimum-width annulus of a set of disks in the plane, described in detail in Section 5. There are also specific applications for the overlay of Voronoi diagrams, for example, representing the local zones of two competing telecommunication operators [39].

5   Minimum-Width Annulus of Disks

In this section we describe an application of our framework that demonstrates its generality and usefulness. We use our tools to compute a minimum-width

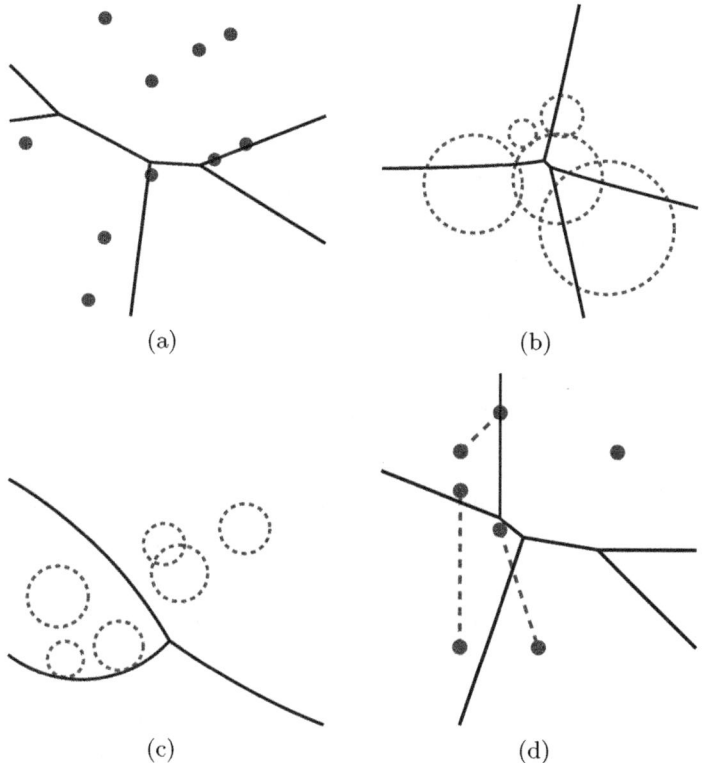

Fig. 8. Farthest-site Voronoi diagrams of sites identical to the sites in Figure 1, computed with our software. (a) A farthest Euclidean Voronoi diagram of points. (b) A farthest Apollonius diagram (additively-weighted Voronoi diagram). (c) A farthest Möbius diagram. (d) A farthest Euclidean Voronoi diagram of segments. The sites in (b), (c), and (d) are illustrated with dashed curves.

annulus containing a set of disks in the plane, by constructing and overlaying two different Voronoi diagrams defined on the input disks.

An *annulus* is the bounded area between two concentric circles. The width of an annulus is the difference between the radius of its outer circle and the radius of its inner circle. Given a set of objects in the plane (disks in our case) the objective is to find a minimum-width annulus (MWA) containing those objects, if one exists.[4] For point sets, a minimum-width annulus can be found by computing the overlay of the nearest-site and farthest-site Voronoi diagrams of the points. The center of the desired annulus must be a vertex of the overlay, and, in fact, an intersection point of the two diagrams [26]. A similar approach handles the case of disks but the diagrams have to be defined more carefully.

---

[4] Generally, when the width of the set is smaller than the width of any containing annulus there is no minimum-width annulus. For example, four collinear points in the plane have no minimum-width annulus.

Consider the following farthest-point distance function ρ from a point $p \in \mathbb{R}^2$ to a set of points $S \subset \mathbb{R}^2$:

$$\rho(p, S) = \sup_{x \in S} ||p - x||,$$

which measures the farthest distance from the point p to the set S. Consider the farthest-site Voronoi diagram with respect to this distance function. We call this diagram the "farthest-point farthest-site" Voronoi diagram. The distance function $\rho(p, S)$ becomes the Euclidean distance when the set S consists of a single point. However, this is not the case when the set S is a disk in the plane, say.

The following observation establishes a connection between the farthest-point farthest-site Voronoi diagram of a set of disks and the Apollonius diagram, and comes in handy below.

Observation 1. *Let $\mathcal{D}$ be a set of disk in $\mathbb{R}^2$. The farthest-point farthest-site Voronoi diagram of $\mathcal{D}$ is identical to the farthest-site Apollonius diagram of the disks with negated radii.*

Indeed, for a disk d with center c and radius r, the farthest-point distance function is $\rho(p, d) = ||p - c|| + r$. On the other hand, the Apollonius distance function is defined to be $\rho(p, d) = ||p - c|| - r$. Hence the farthest-point distance function is identical to the Apollonius distance with "negative radii."

We prove that there is a minimum-width annulus (in case one exists), the center of which is a vertex of the overlay of the Apollonius diagram and the farthest-point farthest-site Voronoi diagram of the disks.

Recall that in the case of point sets the center of a minimum-width annulus is an intersection point of an edge of the nearest-site Voronoi diagram and an edge of the farthest-site Voronoi diagram of the points. In the case of disks, this is not always true. Figure 9 shows a set of disks located at $(2, 0)$, $(0, 3)$, $(-4, 0)$, and $(0, -5)$ with radii 1, 2, 3, and 4, respectively. In this setting, the annulus centered at the origin with radii 1 and 9 is a minimum-width annulus, as its width is equal to a diameter of one of the disks. The origin does not lie on an edge of the farthest-point farthest-site Voronoi diagram, as the outer circle touches only one disk. Nevertheless, the origin is a vertex of the Apollonius diagram, and thus also of the overlay.

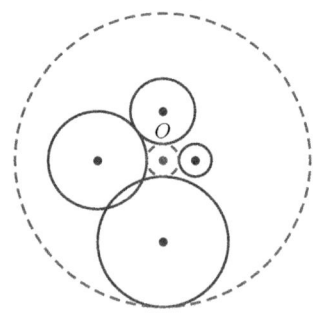

Fig. 9. The outer circle of a MWA that touches one disk

Let $\mathcal{D} = \{d_1, \ldots, d_n\}$ be a collection of disks in the plane such that non of the disks is fully contained in the union of the others, and formally, for all i $d_i \not\subseteq \bigcup_{j \neq i} d_j$. For simplicity of exposition, we assume here that $n \geq 3$; the case of $n < 3$ is simple to handle. We show that there is a minimum-width annulus, the bounding circles of which intersect the disks of $\mathcal{D}$ in at least 4 points, arguing as in [40].

Observation 2. *If N is a minimum-width annulus containing $\mathcal{D}$ then each of the inner circle and the outer circle of N touches a disk of $\mathcal{D}$.*

Indeed, if this is not the case, then we can shrink the outer circle or expand the inner circle of the annulus until each of them touches a disk. Thus, fixing a point p in the plane as the center of a containing annulus fixes an annulus bounding $\mathcal{D}$, of smallest width, for this center. We call this annulus the *tight annulus* fixed at p, denote it as N_p and its width as W_{N_p}. Let $\mathcal{I}_p, \mathcal{O}_p \subseteq \mathcal{D}$ denote the set of disks that are touched by the inner and outer circles of N_p, respectively.

Definition 3 (Neighboring center in direction $\mathbf{d}$). *For a point p and a direction $\mathbf{d}$ in the plane, consider the ray r emanating from p in direction $\mathbf{d}$. r can be divided into maximal contiguous cells, such that every point x in a cell (regarded as the center of a tight annulus) obtains the same $\mathcal{I}_x$ and $\mathcal{O}_x$. Let p' be an interior point of the cell of p, or, if p comprises a single-point cell, an interior point of the neighboring cell in the direction of the ray. We call p' a neighboring center of p in direction $\mathbf{d}$.*

Lemma 1. *There is no minimum-width annulus N_O containing $\mathcal{D}$ and two different disks $A, B \in \mathcal{D}$ such that $\mathcal{I}_O = \{A\}$, $\mathcal{O}_O = \{B\}$, and the centers of A, B and N_O are collinear.*

Proof. Suppose to the contrary that such a minimum-width annulus N_O exists. Denote by O_A and O_B the respective centers of A and B and by R_A and R_B their respective radii. O_A is on the segment OO_B; see the figure to the right for an illustration. Choose O' to be a neighboring center of O in a direction perpendicular to OO_B. Then, by the triangle inequality, $|O_A O_B| > |O'O_B| - |O'O_A|$, and

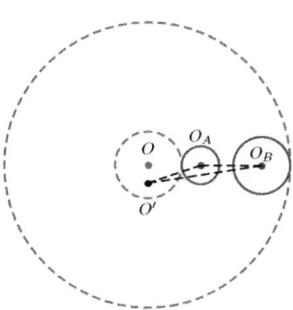

$$W_{N_O} = (|OO_B| + R_B) - (|OO_A| - R_A)$$
$$= |O_A O_B| + R_B + R_A > |O'O_B| - |O'O_A| + R_B + R_A.$$

Thus, $N_{O'}$ has a smaller width, which is a contradiction. □

Lemma 2. *If N_O is a minimum-width annulus containing $\mathcal{D}$ then either $|\mathcal{I}_O| > 1$, or $|\mathcal{O}_O| > 1$, or there is another annulus $N_{O'}$ of minimum-width containing $\mathcal{D}$ such that $|\mathcal{I}_{O'}| > 1$ or $|\mathcal{O}_{O'}| > 1$.*

Proof. Suppose that $\mathcal{I}_O = \{A\}$ and $\mathcal{O}_O = \{B\}$. Let R_A and R_B denote the respective radii of A and B, and let O_A and O_B denote their respective centers.

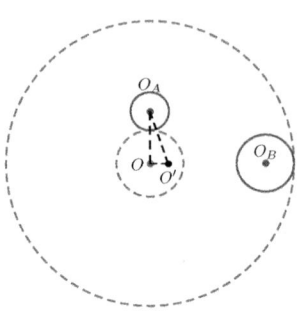

In case that $A = B$, we can move O along the line that goes through O and O_A away from O_A. The new annulus, centered at O', has the same width but the radii of its bounding circles are larger. We keep moving O until one of the circles intersects another disk.

If $A \neq B$ then from Lemma 1, O_A, O_B, and O are not collinear (see the figure on the right for an illustration). Choose O' to be a neighboring center of O in the direction of O_B. Then, by the triangle inequality, $|OO_A| < |OO'| + |O'O_A|$ and

$$W_{N_O} = |OO_B| + R_B - (|OO_A| - R_A) = |OO'| + |O'O_B| + R_B - |OO_A| + R_A$$

$$> |O'O_B| - |O'O_A| + R_B + R_A.$$

Thus the annulus $N_{O'}$ has a smaller width then N_O, which is a contradiction. □

Theorem 2. *If there is a minimum-width annulus containing $\mathcal{D}$, then there is a minimum-width annulus N_O such that $|\mathcal{I}_O| + |\mathcal{O}_O| \geq 4$.*

Proof. Suppose to the contrary that there is a minimum-width annulus N_O containing the disks of $\mathcal{D}$, but there is no minimum-width annulus for which the total number of contact points of its bounding circles with the disks of $\mathcal{D}$ is at least 4.

From Lemma 2, we may assume that either $|\mathcal{I}_O| > 1$, or $|\mathcal{O}_O| > 1$. The only cases that we need to rule out are $|\mathcal{I}_O| = 2$ and $|\mathcal{O}_O| = 1$, or $|\mathcal{I}_O| = 1$ and $|\mathcal{O}_O| = 2$.

Case I: Assume that $\mathcal{I}_O = \{A_1, A_2\}$ and $\mathcal{O}_O = \{B\}$. Let R_{A_1}, R_{A_2}, and R_B denote the respective radii of A_1, A_2, and B, and let O_{A_1}, O_{A_2}, and O_B denote their respective centers.

If $B \in \mathcal{I}_O$ then we can move O along the bisector of A_1 and A_2 (which is one branch of a hyperbola), increasing its distance from A_1 and A_2. W_{N_O} (which equals to $2 \cdot R_B$) will stay the same. As the radii of the bounding circles grows one of the circles will eventually intersect another disk.

If $B \notin \mathcal{I}_O$ then there are two sub-cases:

Case I(a): Neither of O_{A_1} and O_{A_2} lies on the segment OO_B (see the figure to the right for an illustration). Choose O' to be the neighboring center near O in the direction of O_B. Then again from the triangle inequality we get $|OO'| > |OO_{A_i}| - |O'O_{A_i}|$, for $i = 1, 2$, and

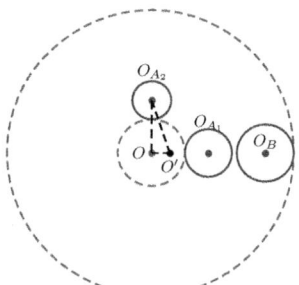

$$
\begin{aligned}
W_{N_O} &= |OO_B| + R_B - (|OO_{A_i}| - R_{A_i}) \\
&= |OO'| + |O'O_B| - |OO_{A_i}| + R_B + R_{A_i} \\
&> |O''O_B| - |O''O_{A_i}| + R_B + R_{A_i} = W^i_{N_O} \ ,
\end{aligned}
$$

for $i = 1, 2$. Since $W_{N_{O''}}$ is the maximum of $W^1_{N_O}$ and $W^2_{N_O}$, it follows that $W_{N_O} > W_{N_{O''}}$, which is a contradiction.

Case I(b): Only one of the points O_{A_1} and O_{A_2} can lie on the segment OO_B, since if both of them lie on the segment then one of the respective disks is fully contained in the other, which contradicts our basic assumption. Without loss of generality, assume that O_{A_1} is on the segment OO_B (see the figure on the right for an illustration). For every point P on the open segment OO_{A_1} we get

$$
\begin{aligned}
|OP| + |PO_{A_2}| - R_{A_2} &> |OO_{A_2}| - R_{A_2} \\
&= |OO_{A_1}| - R_{A_1} = |OP| + |O'O_{A_1}| - R_{A_1},
\end{aligned}
$$

and therefore

$$
|PO_{A_2}| - R_{A_2} > |PO_{A_1}| - R_{A_1}.
$$

This means that neighboring center O' in the direction of O_{A_1} from O satisfies $\mathcal{O}_{O'} = \mathcal{O}_O$ and $\mathcal{I}_{O'} = \mathcal{I}_O \setminus \{A_2\}$. The annulus $N_{O'}$ has the same width as N_O and therefore is also a minimum-width annulus, the center of which is collinear with O_{A_1} and O_B. This, however, is a contradiction to Lemma 1.

Case II: $|\mathcal{I}_O| = 1$ and $|\mathcal{O}_O| = 2$. The proof of this case is similar to the first case and is therefore omitted.   □

From Theorem 2 it is clear that if there is a minimum-width annulus then there is a minimum-width annulus centered at a point O that falls into one of the following three cases:

1. $|\mathcal{I}_O| \geq 3$ and $|\mathcal{O}_O| = 1$. This means that O coincides with a vertex of the Apollonius diagram of the disks.

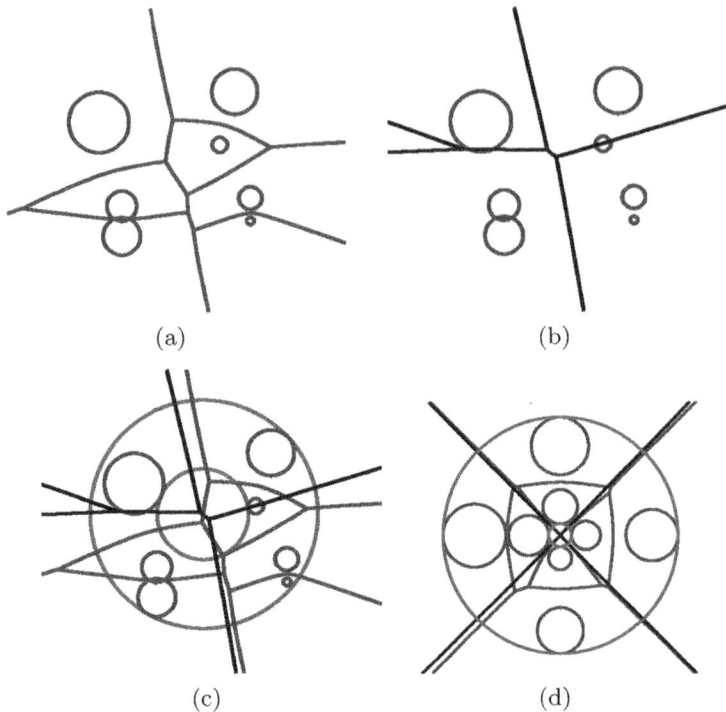

Fig. 10. Computing a minimum-width annulus of a set of disks. (a) The Apollonius diagram of the set of disks. (b) The farthest-point farthest-site Voronoi diagram of the set of disks. (c) A minimum-width annulus of the set of disks. The center of the annulus is a vertex in the overlay of the Apollonius and the farthest-point farthest-site Voronoi diagrams. (d) A highly degenerate scenario for computing a minimum-width annulus of a set of disks. Four disks induce the vertex of the farthest-point farthest-site Voronoi diagram and other four disks induce a vertex of the Apollonius diagram at the same point.

2. $|\mathcal{I}_O| = 1$ and $|\mathcal{O}_O| \geq 3$. This means that O coincides with a vertex of the farthest-point farthest-site Voronoi diagram of the disks.
3. $|\mathcal{I}_O| \geq 2$ and $|\mathcal{O}_O| \geq 2$. In this case O lies on a Voronoi edge of the Apollonius diagram of the disks and on a Voronoi edge of the farthest-point farthest-site Voronoi diagram of the disks.

We therefore construct each of the diagrams (the Apollonius and the farthest-point farthest-site Voronoi diagrams), using our general divide-and-conquer machinery, and then overlay them. For each vertex O of the overlay, we retrieve four relevant disks (either three touching the inner circle and the one touching the outer circle, in case 1, or three touching the outer circle and the one touching the inner circle, in case 2, or two pairs of disks touching respectively the inner and outer circles, in case 3), and compute the width of the resulting annulus for those four disks. We output the annulus of the smallest width. Figures 10(a)–(c) illustrate various steps of the algorithm for computing a minimum-width annulus of a set of disks. Figure 10(d) depicts a highly degenerate input that is being handled properly by our implementation.

Apollonius diagrams are of linear complexity in the number of sites and conform to the *abstract Voronoi diagram* definition [6]. Hence, farthest-site Apollonius diagrams are *farthest abstract Voronoi diagrams* [41], and are, therefore, of linear complexity too. Thus Corollary 1 is applied to yield an expected construction time of $O(n \log^2 n)$ for both diagrams. Overlaying the two diagrams with the sweep-based algorithm has $O((n + k) \log n)$ worst-case time complexity where k is the number of intersections between edges of the diagrams. The total expected running time of the algorithm is therefore $O(n \log^2 n + k \log n)$, where k can be[5] $\Theta(n^2)$.

The implementation of the Apollonius diagram is based on the algebraic curves traits for the arrangement package [27,28], as the bisector of two sites is, in general, one branch of a hyperbola. In order to determine which of the branches of the hyperbola constitutes the bisector of the two disks, we first lexicographically sort the x-monotone sub-curves of the hyperbola. Then, by considering the coordinates of the centers of the disks and their radii, we determine which of the two branches is the bisector. For example, in the case of a vertical asymptote, if the center of the disk of the smaller radius is to the left of the center of the disk of the larger radius, the desired branch is the left branch of the hyperbola, otherwise the desired branch is the right branch.

Table 2. Time consumption (in seconds) of minimum-width annuli computation and sizes of the corresponding constructed diagrams. FPFS — farthest-point farthest-site, V, E, F — the number of vertices, edges, and faces of the diagram, T — the time in seconds consumed by the respective phase, C — the time in seconds consumed during the comparison of the candidates for the annulus center. "*dgn_**" are input data in a degenerate setting (see Figure 10(d) for an illustration), and "*rnd_**" are input data in a random setting.

Input	Apollonius				FPFS VD				Overlay				C	Total Time
	V	E	F	T	V	E	F	T	V	E	F	T		
rnd_50	84	128	45	6.74	10	21	12	2.40	126	213	88	0.44	0.57	10.16
rnd_100	162	242	81	17.47	17	35	19	5.36	238	395	158	0.81	1.46	25.12
rnd_200	317	460	144	44.07	15	31	17	11.16	416	659	244	1.28	1.33	57.86
rnd_500	672	967	296	136.46	16	33	18	29.30	775	1174	400	1.85	1.97	169.59
dgn_50	59	106	48	5.74	1	25	25	1.94	100	213	114	0.84	0.30	8.83
dgn_100	134	232	99	16.57	1	50	50	5.83	239	492	254	2.93	1.11	26.46
dgn_200	304	502	199	42.60	1	100	100	15.77	581	1156	576	7.60	2.78	68.76
dgn_500	785	1262	478	154.37	1	239	239	56.21	1645	3221	1577	37.93	7.94	256.47

---

[5] It is reasonable to assume that the expected time complexity is, in many cases, smaller. Though not proven for disks, it is known that, for random point sets the expected number of intersections between the farthest-site and the nearest-site Voronoi diagrams is sub-linear [42].

Given this implementation of the Apollonius traits class, the general framework allows us to easily compute the farthest-point farthest-site Voronoi diagram by computing the farthest Voronoi diagram of the disks with "negative radii." The overlay operation is readily available in the arrangement package.

Table 2 shows the sizes of the constructed diagrams in each phase of the algorithm and the time consumption (in seconds) of the execution of each phase. The vertices of the overlay are the candidates for the center of the annulus. The experiments were carried out on an Intel® Core™2 Duo 2GHz processor with 1GB memory running Linux operating system.

6   Conclusions and Future Work

Our framework together with the existing traits classes of the arrangement package provide the means to produce various Voronoi diagrams in an exact and robust manner. Moreover, the theoretical bound on the expected running time of our algorithm is nearly optimal.

Future work is to enrich the variety of Voronoi diagrams that can be computed with our software to include Voronoi diagrams of circular arcs, Bregman Voronoi diagrams [43], Voronoi diagrams in the hyperbolic Poincaré half-plane [44] or in the Poincaré hyperbolic disk [45], Hausdorff Voronoi Diagrams [2], and Voronoi diagrams on different surfaces, for example, cylinders or tori. Some of the above diagrams have bisectors that can be handled with existing traits classes available by the arrangement package. Another objective is to generalize the computation of minimum-width annuli to other types of objects, for example, ellipses.

Acknowledgements. The authors are grateful to Efi Fogel and Eric Berberich for many helpful suggestions and fruitful cooperation, and to Ron Wein, Michal Meyerovitch, and Barak Raveh for their cooperation and support of this work.

References

1. Shamos, M.I., Hoey, D.: Closest-point problems. In: Proceedings of the 16th IEEE Symposium on the Foundations of Computer Science, pp. 151–162 (1975)
2. Okabe, A., Boots, B., Sugihara, K., Chiu, S.N.: Spatial Tessellations: Concepts and Applications of Voronoi Diagrams, 2nd edn. John Wiley & Sons, NYC (2000)
3. Aurenhammer, F., Klein, R.: Voronoi diagrams. In: Sack, J.R., Urrutia, J.B. (eds.) Handbook of Computational Geometry, pp. 201–290. Elsevier Science Publishers, B.V., North-Holland (2000)
4. Boissonnat, J.D., Wormser, C., Yvinec, M.: Curved Voronoi diagrams. In: Boissonnat, J.D., Teillaud, M. (eds.) Effective Computational Geometry for Curves and Surfaces. Springer, Heidelberg (2006)
5. Barequet, G., Dickerson, M.T., Drysdale, R.L.S.: 2-point site Voronoi diagrams. Discrete Applied Mathematics 122(1-3), 37–54 (2002)
6. Klein, R.: Concrete and Abstract Voronoi Diagrams. LNCS, vol. 400. Springer, Heidelberg (1989)

7. Fortune, S.J.: A sweepline algorithm for Voronoi diagrams. Algorithmica 2(1-4), 153–174 (1987)
8. Guibas, L.J., Knuth, D.E., Sharir, M.: Randomized incremental construction of Delaunay and Voronoi diagrams. Algorithmica 7(1-6), 381–413 (1992)
9. Klein, R., Mehlhorn, K., Meiser, S.: Randomized incremental construction of abstract Voronoi diagrams. Computational Geometry: Theory and Applications 3(3), 157–184 (1993)
10. Karavelas, M.I., Yvinec, M.: The Voronoi diagram of planar convex objects. In: Di Battista, G., Zwick, U. (eds.) ESA 2003. LNCS, vol. 2832, pp. 337–348. Springer, Heidelberg (2003)
11. Aichholzer, O., Aigner, W., Aurenhammer, F., Hackl, T., Jüttler, B., Pilgerstorfer, E., Rabl, M.: Divide-and-conquer for Voronoi diagrams revisited. In: Proceedings of the 25th Annual ACM Symposium on Computational Geometry (SoCG), pp. 189–197. Association for Computing Machinery (ACM) Press, New York (2009)
12. Boissonnat, J.D., Devillers, O., Pion, S., Teillaud, M., Yvinec, M.: Triangulations in CGAL. Computational Geometry: Theory and Applications 22(1-3), 5–19 (2002)
13. Emiris, I.Z., Karavelas, M.I.: The predicates of the Apollonius diagram: Algorithmic analysis and implementation. Computational Geometry: Theory and Applications 33(1-2), 18–57 (2006)
14. Karavelas, M.I.: A robust and efficient implementation for the segment Voronoi diagram. In: Proceedings of the 1st International Symposium on Voronoi Diagrams in Science and Engineering (ISVD), pp. 51–62 (2004)
15. Emiris, I.Z., Tsigaridas, E.P., Tzoumas, G.: Voronoi diagram of ellipses in CGAL. In: Abstracts of the 24th European Workshop on Computational Geometry, pp. 87–90 (2008)
16. Burnikel, C., Mehlhorn, K., Schirra, S.: How to compute the Voronoi diagram of line segments: Theoretical and experimental results. In: van Leeuwen, J. (ed.) ESA 1994. LNCS, vol. 855, pp. 227–239. Springer, Heidelberg (1994)
17. Held, M.: VRONI: An engineering approach to the reliable and efficient computation of Voronoi diagrams of points and line segments. Computational Geometry: Theory and Applications 18(2), 95–123 (2001)
18. Hoff III, K.E., Keyser, J., Lin, M., Manocha, D., Culver, T.: Fast computation of generalized Voronoi diagrams using graphics hardware. In: Proceedings of the 26th Annual International Conference on Computer Graphics and Interactive Techniques, pp. 277–286. Association for Computing Machinery (ACM) Press, New York (1999)
19. Nielsen, F.: An interactive tour of Voronoi diagrams on the GPU. In: ShaderX[6]: Advanced Rendering Techniques. Charles River Media, Hingham (2008)
20. Edelsbrunner, H., Seidel, R.: Voronoi diagrams and arrangements. Discrete & Computational Geometry 1(1), 25–44 (1986)
21. Meyerovitch, M.: Robust, generic and efficient construction of envelopes of surfaces in three-dimensional space. In: Azar, Y., Erlebach, T. (eds.) ESA 2006. LNCS, vol. 4168, pp. 792–803. Springer, Heidelberg (2006)
22. Meyerovitch, M., Wein, R., Zukerman, B.: 3D Envelopes. In: CGAL Editorial Board (ed.) CGAL User and Reference Manual, 3.4th edn. (2008)
23. Agarwal, P.K., Schwarzkopf, O., Sharir, M.: The overlay of lower envelopes and its applications. Discrete & Computational Geometry 15(1), 1–13 (1996)
24. Setter, O.: Constructing two-dimensional Voronoi diagrams via divide-and-conquer of envelopes in space. M.Sc. thesis, The Blavatnik School of Computer Science, Tel-Aviv University, Tel-Aviv, Israel (May 2009), Electronic version can be found at http://arxiv.org/abs/0906.2760

25. Berberich, E., Fogel, E., Halperin, D., Mehlhorn, K., Wein, R.: Arrangements on parametric surfaces I: General framework and infrastructure (2010); Accepted for publication in Mathematics in Computer Science
26. Ebara, H., Fukuyama, N., Nakano, H., Nakanishi, Y.: Roundness algorithms using the Voronoi diagrams. In: Abstracts of the 1st Canadian Conference on Computational Geometry, p. 41 (1989)
27. Eigenwillig, A., Kerber, M.: Exact and efficient 2D-arrangements of arbitrary algebraic curves. In: Proceedings of the 19th Annual ACM-SIAM Symposium on Discrete Algorithms (SODA), pp. 122–131. Society for Industrial and Applied Mathematics (SIAM), Philadelphia (2008)
28. Berberich, E., Emeliyanenko, P.: CGAL's curved kernel via analysis. ACS Technical Report ACS-TR-123203-04, MPI (2008)
29. Sugihara, K.: Laguerre Voronoi diagram on the sphere. Journal for Geometry and Graphics 6(1), 69–81 (2002)
30. Aurenhammer, F., Edelsbrunner, H.: An optimal algorithm for constructing the weighted Voronoi diagram in the plane. Pattern Recognition 17(2), 251–257 (1984)
31. Yap, C.K., Dubé, T.: The exact computation paradigm. In: Du, D.Z., Hwang, F.K. (eds.) Computing in Euclidean Geometry, 2nd edn. LNCS, vol. 1, pp. 452–492. World Scientific, Singapore (1995)
32. Austern, M.H.: Generic Programming and the STL. Addison-Wesley, Reading (1999)
33. Wein, R., Fogel, E., Zukerman, B., Halperin, D.: 2D arrangements. In: CGAL Editorial Board (ed.) CGAL User and Reference Manual, 3.4th edn. (2008)
34. Sharir, M.: The Clarkson-Shor technique revisited and extended. Combinatorics, Probability & Computing 12(2) (2003)
35. Guibas, L.J., Seidel, R.: Computing convolutions by reciprocal search. Discrete & Computational Geometry 2(1), 175–193 (1987)
36. Berberich, E., Fogel, E., Halperin, D., Kerber, M., Setter, O.: Arrangements on parametric surfaces II: Concretization and applications (2010); Accepted for publication in Mathematics in Computer Science
37. Fogel, E., Setter, O., Halperin, D.: Movie: Arrangements of geodesic arcs on the sphere. In: Proceedings of the 24th Annual ACM Symposium on Computational Geometry (SoCG), pp. 218–219. Association for Computing Machinery (ACM) Press, New York (2008)
38. Knuth, D.E.: Art of Computer Programming, 2nd edn. Sorting and Searching, vol. 3. Addison-Wesley, Reading (April 1998)
39. Baccelli, F., Gloaguen, C., Zuyev, S.: Superposition of planar Voronoi tessellations. Stochastic Models 16, 69–98 (2000)
40. Rivlin, T.J.: Approximating by circles. Computing 21, 93–104 (1979)
41. Mehlhorn, K., Meiser, S., Rasch, R.: Furthest site abstract Voronoi diagrams. International Journal of Computational Geometry and Applications 11(6), 583–616 (2001)
42. Bose, P., Devroye, L.: Intersections with random geometric objects. Computational Geometry: Theory and Applications 10(3), 139–154 (1998)
43. Nielsen, F., Boissonnat, J.D., Nock, R.: On Bregman Voronoi diagrams. In: Proceedings of the 18th Annual ACM-SIAM Symposium on Discrete Algorithms (SODA), pp. 746–755. Society for Industrial and Applied Mathematics (SIAM), Philadelphia (2007)
44. Onishi, K., Takayama, N.: Construction of Voronoi diagram on the upper half-plane. IEICE Transactions on Fundamentals of Electronics, Communications and Computer Sciences 79(4), 533–539 (1996)

45. Nilforoushan, Z., Mohades, A.: Hyperbolic Voronoi diagram. In: Gavrilova, M.L., Gervasi, O., Kumar, V., Tan, C.J.K., Taniar, D., Laganá, A., Mun, Y., Choo, H. (eds.) ICCSA 2006. LNCS, vol. 3984, pp. 735–742. Springer, Heidelberg (2006)
46. Stroustrup, B.: The C++ Programming Language, 3rd edn. Addison-Wesley, Boston (1997)
47. Myers, N.: "Traits": A new and useful template technique. In: Lippman, S.B. (ed.) C++ Gems. SIGS Reference Library, vol. 5, pp. 451–458 (1997)
48. Berberich, E., Kerber, M.: Exact arrangements on tori and Dupin cyclides. In: Proceedings of the 2008 ACM Symposium on Solid and Physical Modeling (SPM), pp. 59–66. Association for Computing Machinery (ACM) Press, New York (2008)

A   Software Interface: Adding New Diagrams

This appendix contains technical details on what needs to be supplied by the user of the framework in order to implement a new type of Voronoi diagrams. The interface consists of several functions, each operating on a small number of user-defined types (sites or bisector curves). Providing these functions does not require knowledge of the underlying algorithms. In this appendix we assume some familiarity of the reader with generic programming [32] and the C++ programming-language [46].

The Envelope_3 package is fairly general and can deal with surfaces that have two-dimensional intersections [22]. Therefore, the Envelope_3 can also compute Voronoi diagrams with two-dimensional bisectors.[6] However, most types of two-dimensional Voronoi diagrams have only one-dimensional bisectors. We used this fact to reduce and simplify the interface that is used by the code for computing Voronoi diagrams with one-dimensional bisectors via envelopes. In order to construct Voronoi diagrams with two-dimensional bisectors we refer the user to [24].

Generic programming [32] manifests a formal hierarchy of polymorphic abstract requirements on data types referred to as *concepts*, and a set of classes that conform precisely to the specified requirements, referred to as *models*. Models that describe behaviors are referred to as *traits classes* [47]. Following the generic-programming approach, we implemented a generic template-function called voronoi_2<GeometryTraits,TopologyTraits> that computes Voronoi diagrams via envelope constructions.

When the function-template is instantiated, the GeometryTraits parameter must be substituted with a geometry-traits class — a model of the concept *EnvelopeVoronoiTraits_2*, and the TopologyTraits parameter must be substituted with a topology-traits class. The latter adapts the underlying data-structure that maintains the incidence relations on the arrangement features to the embedding surface, which is parameterized by two parameters u and v. (In case of the plane, for example, u and v map to x and y, respectively.) Thus, the developer of a new diagram type must choose a topology-traits class that handles the

---

[6] For example, two point sites in the L_1-Voronoi diagram can have a two-dimensional bisector.

embedding surface of the desired diagram; for more information on topology-traits and the surfaces they support refer to [25]. The `Arrangement_on_surface_2` package currently includes topology-traits classes that handle the bounded or unbounded plane, topology-traits classes that handle elliptic quadrics and ring Dupin cyclides that generalize tori [48], and a specially tailored topology-traits class that handles the sphere.

The `Arrangement_on_surface_2` package supports robust construction of arrangements embedded on two-dimensional parametric surfaces in 3D. We use such arrangements to represent Voronoi diagrams embedded on corresponding surfaces. The `Arrangement_on_surface_2` package defines the *ArrangementTraits_2* concept, the models of which are geometry-traits classes that contain geometric primitives used by the package. The *EnvelopeVoronoiTraits_2* concept refines the *ArrangementTraits_2* concept. The predicates and operations defined by the *ArrangementTraits_2* are necessary for the manipulation of bisector curves of Voronoi sites.

The *EnvelopeVoronoiTraits_2* concept defines additional types and operations on top of those defined by the *ArrangementTraits_2* concept. The `Site_2` type represents a Voronoi site. Given two variables of type `Site_2` and an output iterator, the `Construct_bisector_2` functor[7] returns a sequence of objects of type `X_monotone_curve_2` that together form the bisector of the two Voronoi sites. If the bisector between the two sites does not exist, the function returns an empty sequence.[8]

Other required functors are used to determine which side of the bisector, if there is one, each site dominates. The `Compare_distance_above_2` functor accepts two site objects and a u-monotone curve, which is part of their bisector, and indicates which site dominates the region above the u-monotone curve, where "above" is defined to be the region to the left of the u-monotone curve when it is traversed from the uv-lexicographically smaller end to the uv-lexicographically larger end. The framework utilizes the fact that each of the sites dominates one side of the bisector to implement the "below" version of the functor. If there is no bisector between the two sites then the `Compare_dominance_2` functor is used to indicate which of the sites dominates the whole domain.

Given two input sites and a point, the `Compare_distance_at_point_2` functor indicates which site dominates the point in the two-dimensional domain. The functor is used together with the `Construct_point_on_x_monotone_curve_2` functor that constructs an interior point on a given u-monotone curve.

The `voronoi_2` function uses the `Envelope_3` package to compute the desired Voronoi diagram. It uses the class template `Voronoi_2_to_Envelope_3_adaptor` to adapt models of *EnvelopeVoronoiTraits_2* concept to classes that conform to the

---

[7] Functor is a synonym for "function object" and is invoked or called as if it were an ordinary function with the same syntax.

[8] There are cases where there is no bisector between two Voronoi sites. For example, two Apollonius sites where one is completely contained inside the other have no bisector.

requirements of the `Envelope_3` package. A similar function `farthest_voronoi_2` is used to compute farthest-site Voronoi diagrams.

Table 3 summarizes the types and functors that are needed in order to implement a new Voronoi diagram type.

Table 3. Required types and functors by the *EnvelopeVoronoiTraits_2* concept

Name	*Input*	*Output*	*Description*
Site_2	—	—	A type that represents a Voronoi site.
Construct_-bisector_2	Two Site_2 objects	An output iterator with values of type of X_monotone_-curve_2	Returns X_monotone_curve_2 objects that together form the bisector of the two input sites.
Compare_-distance_-above_2	Two Site_2 objects and an X_mono-tone_curve_2	Comparison_result	Determines which of the given Voronoi sites is closer to the area "above" the given *u*-monotone curve, where "above" is the area that lies to its left when the curve is traversed from its *uv*-lexicographically smaller end to its *uv*-lexicographically larger end.
Compare_-distance_at_-point_2	Two Site_2 objects and a Point_2	Comparison_result	Determines which of the given Voronoi sites is closer to the given point.
Compare_-dominance_2	Two Site_2 objects	Comparison_result	Determines which of the sites dominates the domain in case that there is no bisector between the two sites.
Construct_-point_on_-x_monotone_-curve_2	X_monotone_-curve_2	Point_2	Constructs an interior point on the given *u*-monotone curve.

Approximate Shortest Path Queries Using Voronoi Duals

Shinichi Honiden, Michael E. Houle, Christian Sommer⋆, and Martin Wolff

National Institute of Informatics, Tokyo, Japan
sommer@is.s.u-tokyo.ac.jp

Abstract. We propose an approximation method to answer point-to-point shortest path queries in undirected edge-weighted graphs, based on random sampling and Voronoi duals. We compute a simplification of the graph by selecting nodes independently at random with probability p. Edges are generated as the Voronoi dual of the original graph, using the selected nodes as Voronoi sites. This overlay graph allows for fast computation of approximate shortest paths for general, undirected graphs. The time–quality tradeoff decision can be made at query time. We provide bounds on the approximation ratio of the path lengths as well as experimental results. The theoretical worst-case approximation ratio is bounded by a logarithmic factor. Experiments show that our approximation method based on Voronoi duals has extremely fast preprocessing time and efficiently computes reasonably short paths.

1 Introduction

We wish to answer shortest path queries for large edge-weighted graphs such as those stemming from transportation networks, social networks, protein interaction networks, and the web graph. One could use a classical single source shortest path algorithm such as Dijkstra's [Dij59], which has worst-case running time $O(m + n \lg n)$, where n denotes the number of nodes and m the number of edges. However, for large graphs, only a relatively small portion of the graph can be searched at query time. If preprocessing is allowed, queries can be answered much more quickly. The algorithms with fastest query times are those that precompute and store the shortest-path distances between all possible pairs of source and target — that is, those that precompute solutions to the All Pairs Shortest Path Problem. Shortest path queries could then be answered in constant time. The fastest known algorithm for computing all shortest paths runs in time $O(n^3 / \lg^2 n)$ [Cha07]. Unfortunately, the preprocessing time is prohibitively large in practice.

The goal is to mediate between the two extremes of no precomputation of paths and total precomputation of paths. The desired tradeoff between preprocessing time and query time depends on the needs of the application.

⋆ Corresponding Author.

M.L. Gavrilova et al. (Eds.): Trans. on Comput. Sci. IX, LNCS 6290, pp. 28–53, 2010.

1.1   Related Work

In the following, we give a brief overview of related work for shortest path and distance queries.

Theoretical. Data structures allowing for shortest path or distance queries are referred to as distance oracles. Their construction is closely related to that of graph spanners. For a pair of nodes (s, t), an approximate distance oracle is said to have stretch (α, β) if it returns a distance in the range $[d(s,t), \alpha \cdot d(s,t) + \beta]$ [EP04]. A girth conjecture by Erdős implies that, for general undirected graphs, distance oracles with multiplicative stretch $\alpha < 2k + 1$ need $\Omega(n^{1+1/k})$ space. An algorithm by Thorup and Zwick [TZ05] constructs such an oracle in expected time $O(kmn^{1/k})$ with query time $O(k)$ and stretch $(2k - 1, 0)$. For constant k, except for the preprocessing time, all their bounds are essentially tight. Baswana and Kavitha [BK06] provide a solution with preprocessing time of $O(n^2 \lg n)$. For unweighted graphs, subquadratic preprocessing time is possible [BGSU08]. For planar (directed) graphs with integer weights, an algorithm by Thorup [Tho04a] constructs a $(1 + \epsilon, 0)$-stretch oracle in time $O(n \lg^3 n \lg(n\Delta))$, where Δ denotes the largest weight. Unfortunately, for huge non-planar graphs, these results are not practical.

Practical. The main focus of practical investigations so far has been on large road networks. There has been considerable recent progress: for the road networks of Europe or the USA, using a high-performance computer, a speed-up of several orders of magnitude compared to Dijkstra's algorithm can be achieved with a preprocessing time in the tens of minutes [SS07a]. Unfortunately, theoretical bounds on both query time and preprocessing time are often difficult to obtain. Goldberg and Harrelson [GH05] proposed a variant of A* search [Dor67] in which distances are precomputed with respect to a small set of 'landmark' vertices. Hierarchical methods [GSSD08, SS07b] provide an efficient framework, especially in the case of road networks. Sanders and Schultes [SS07a, SS07b, SS06] developed a method to compute shortest paths in 'almost constant time' with a carefully designed structure consisting of precomputed shortest paths. Their solution is tailored to perform exceptionally well for road networks, where graphs are almost planar and nodes have small constant degrees. Precomputation is time- and space-consuming; however, it is still manageable in practice, and allows for extremely fast query times.

Even though road networks constitute the most common and popular application of shortest path query algorithms to date, other challenging applications exist. Computer networks, social networks, protein interaction networks, and the web graph exhibit different degree and structural properties, and may contain hundreds of millions or even billions of nodes. In specific cases, a user might be willing to trade preprocessing time against exactness due to the vast size of the data or due to restricted processing power. These scenarios may require the use of a fast approximation method.

1.2   Contribution

We propose an approximation method to answer shortest path queries in general, undirected graphs with positive edge weights, based on random sampling and graph Voronoi duals [Meh88, Erw00]. In preprocessing, each node is selected as a Voronoi site independently at random with probability p, and the Voronoi dual is computed for the selected sites (Section 3). This preprocessing step is very efficient; it takes time proportional to computing one single source shortest path tree (Section 4). For $p < 1$, the resulting dual graph is expected to be smaller than the original graph. At query time, search for the shortest path from source s to target t can potentially be done faster in the Voronoi dual. We let the shortest path in the Voronoi dual guide the search for an approximate shortest path in the original graph. We prove that the expected approximation ratio is at most logarithmic in the number of nodes on the actual shortest path, and that this bound is tight (Section 5). Our experimental results show that, in practice, the approximation is much better than the stated theoretical bound and that the preprocessing overhead is indeed extremely low (Section 6).

2   Preliminaries

An *edge-weighted* graph $G = (V, E, \omega)$ consists of a graph (V, E) together with a *weight function* $\omega : E \to \mathbb{R}$. We assume positive edge weights; that is, $\omega : E \to \mathbb{R}^+$. For the remainder of the paper, we will refer to the number of nodes and edges of the graph by $n = |V|$ and $m = |E|$, respectively.

A *path* from $s = u_0 \in V$ to $t = u_h \in V$ is a node sequence $(u_0, u_1, \ldots, u_h)$ for which $(u_i, u_{i+1}) \in E$ for all $i \in \{0, 1, \ldots h - 1\}$. The *length* of a path P is the sum of its edge weights $\ell(P) := \sum_{i=0}^{h-1} \omega(u_i, u_{i+1})$. A subpath P' of a path P is a subsequence of its nodes $P' = (u_i, u_{i+1}, \ldots u_j)$, $0 \le i < j \le h$. A *simple path* is a path without repeated vertices. Let $\mathcal{P}_G(u, v)$ denote the set of paths from u to v in G. The *distance* $d(u, v)$ between two nodes u, v is the length of a shortest path from u to v; that is, $d(u, v) = \min_{P \in \mathcal{P}(u,v)} \ell(P)$. If $\mathcal{P}(u, v) = \emptyset$ then $d(u, v) := \infty$. Let $SP_G(s, t)$ be an arbitrary shortest path from s to t. Analogously to the multiplicative stretch of a distance oracle, we define the *stretch* of a path P from s to $t \ne s$ as the ratio $\ell(P)/\ell(SP_G(s, t))$.

2.1   Graph Voronoi Diagram

The classical Voronoi diagram is a distance-based decomposition of a metric space relative to a discrete set, the Voronoi sites [Dir50, Vor07]. For a survey on this fundamental structure, we refer to [Aur91]. Among many applications, the Voronoi diagram is often used to solve facility location problems [Sha75, ACS99, AGK+04, GKP05, Svi08]. The Voronoi diagram and the Delaunay triangulation of n points in the plane can be computed in expected time $n \cdot 2^{O(\sqrt{\lg \lg n})}$ [CP07], which is even faster than $O(n \lg n)$.

Mehlhorn [Meh88] and Erwig [Erw00] proposed an analogous decomposition, the *Graph Voronoi Diagram*, for undirected and directed graphs respectively.

Since the Voronoi diagram for the Euclidean space is used for various applications, its graph counterpart, the graph Voronoi diagram, may be used for these applications if the underlying metric is the shortest path metric of a graph. Real-world distances or travelling times can be approximated more appropriately using models based on weighted graphs. In general, non-planar networks such as social networks, computer networks, protein interaction networks, and the web graph cannot be embedded into a low-dimensional Euclidean space without significant distortion.

Definition 1 (Graph Voronoi Diagram [Meh88, Erw00]). *In a graph $G = (V, E, \omega)$, the Voronoi diagram for a set of nodes $K = \{v_1, \ldots, v_k\} \subseteq V$ is a disjoint partition $\mathsf{Vor}_{(G,K)} := \{V_1, \ldots, V_k\}$ of V such that for each node $u \in V_i$, $d(u, v_i) \leq d(u, v_j)$ for all $j \in \{1, \ldots, k\}$.*

The V_i are called *Voronoi regions*. The graph Voronoi diagram is not necessarily unique, as a node u may have the same distance to more than one Voronoi node. Let $\mathsf{vor}(u)$ denote the index i of the Voronoi region V_i containing u; that is, $\mathsf{vor}(u) = i \Leftrightarrow u \in V_i$.

Analogously to the Delaunay triangulation dual for classical Voronoi diagrams of point sets, we define the Voronoi dual for graphs.

Definition 2. *Let $G = (V, E, \omega)$ be an edge-weighted graph and $\mathsf{Vor}_{G,K}$ its Voronoi diagram. The Voronoi dual is the graph $G^* = (K, E^*, \omega^*)$ with edge set $E^* := \{(v_i, v_j) : v_i, v_j \in K \text{ and } \exists u \in V_i \wedge \exists w \in V_j : (u, w) \in E\}$, and edge weights $\omega^*(v_i, v_j) := \min\limits_{\substack{u \in V_i, w \in V_j \\ (u,w) \in E}} \{d(v_i, u) + \omega(u, w) + d(w, v_j)\}$.*

Figure 1 illustrates two graph Voronoi diagrams for the same (planar) graph but with different edge weights. Although the classical Voronoi dual of a non-degenerate set of points in the plane is always a triangulation, the graph Voronoi dual is not necessarily a triangulation, even for planar graphs. For example, a graph Voronoi dual may have nodes whose removal would disconnect the graph.

Erwig [Erw00, Theorem 2] showed that the graph Voronoi diagram can be constructed with a single Dijkstra search in time $O(m + n \cdot \lg n)$. A heap is used to store the shortest path distances from nodes to their closest Voronoi node. The heap is initialized to store the Voronoi nodes themselves. Thereafter, as long as there are nodes in the queue, the minimum is extracted from the heap and processed (or 'settled') by assigning to it a Voronoi region, storing the distance to its Voronoi node, and adding to or updating its neighbors in the queue. We slightly modify this construction of the Voronoi *diagram* [Erw00, Section 3.1] to compute the Voronoi *dual* — that is, to also compute E^* and ω^*. Whenever a node u is settled in the Dijkstra search, for all its settled neighbors u' of different Voronoi regions, the edge $(v^*_{\mathsf{vor}(u)}, v^*_{\mathsf{vor}(u')})$ with weight $\omega_{G^*}(v^*_{\mathsf{vor}(u)}, v^*_{\mathsf{vor}(u')}) = d_G(v_{\mathsf{vor}(u)}, u) + \omega_G(u, u') + d_G(u', v_{\mathsf{vor}(u')})$ is added, or its length is decreased if there already is an edge in G^* representing a longer path in G. This modification of Erwig's algorithm is shown as Algorithm 1.

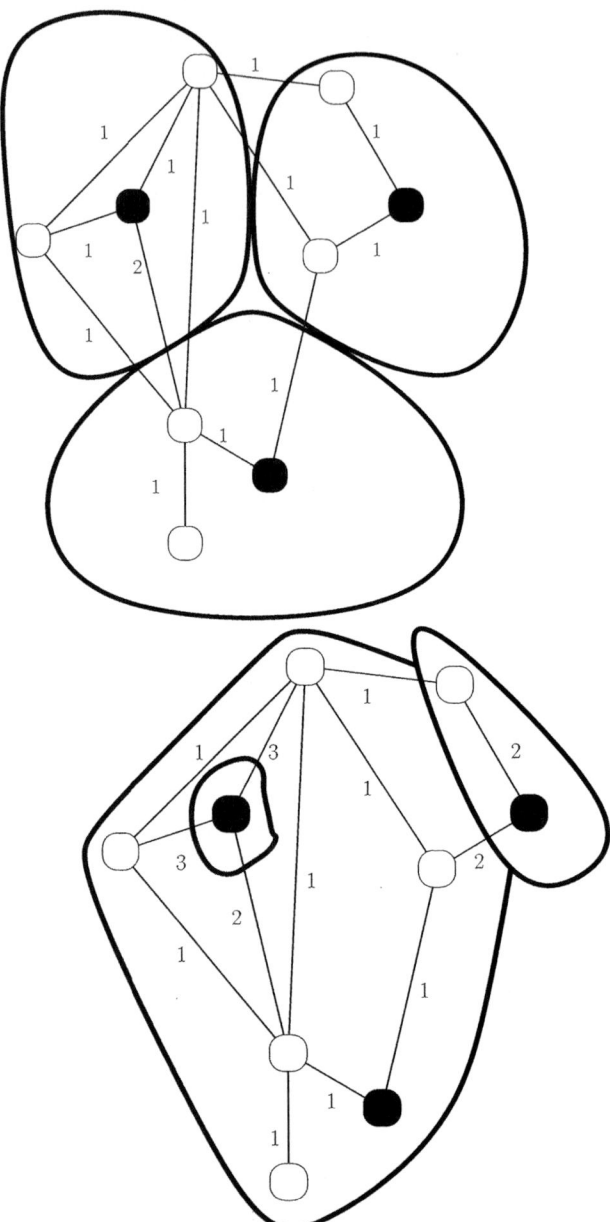

Fig. 1. Two graph Voronoi diagrams for the same planar graph but with different edge weights. Voronoi nodes are black and the remaining nodes are white. Even though the graphs are structurally equivalent, the corresponding graph Voronoi diagrams are not.

Algorithm 1. ComputeVoronoiDual($G = (V, E), K \subseteq V$)

1: **for** $i := 1$ **to** $k = |K|$ **do**
2:     $\text{vor}(v_i) := i$
3:     HEAP.put(v_i)
4: **end for**
5: **while** $\neg$HEAP.empty **do**
6:     $u_{\text{cur}} :=$ HEAP.extractMin
7:     **for** $u \in \Gamma(u_{\text{cur}})$ **do**
8:         **if** $\text{vor}(u) = undefined$ **then**
9:             $\text{vor}(u) := \text{vor}(u_{\text{cur}})$
10:             HEAP.insert($u, d(v_d, u_{\text{cur}}) + \omega(u_{\text{cur}}, u)$)
11:         **else if** $d(v_d, u_{\text{cur}}) + \omega(u_{\text{cur}}, u) < d(v_d, u)$ **then**
12:             $\text{vor}(u) := \text{vor}(u_{\text{cur}})$
13:             HEAP.decreaseKey($u, d(v_d, u_{\text{cur}}) + \omega(u_{\text{cur}}, u)$)
14:         **else if** $\neg$HEAP.contains(u) **and** $\text{vor}(u) \neq \text{vor}(u_{\text{cur}})$ **then**
15:             **if** $\left(v_{\text{vor}(u_{\text{cur}})}, v_{\text{vor}(u)}\right) \notin E^*$ **then**
16:                 $E^* := E^* \cup \left\{\left(v_{\text{vor}(u_{\text{cur}})}, v_{\text{vor}(u)}\right)\right\}$
17:                 $\omega^*\left(v_{\text{vor}(u_{\text{cur}})}, v_{\text{vor}(u)}\right) := \infty$
18:             **end if**
19:             **if** $\omega^*\left(v_{\text{vor}(u_{\text{cur}})}, v_{\text{vor}(u)}\right) > d(v_{\text{vor}(u_{\text{cur}})}, u_{\text{cur}}) + \omega(u_{\text{cur}}, u) + d(u, v_{\text{vor}(u)})$ **then**
20:                 $\omega^*\left(v_{\text{vor}(u_{\text{cur}})}, v_{\text{vor}(u)}\right) := d(v_{\text{vor}(u_{\text{cur}})}, u_{\text{cur}}) + \omega(u_{\text{cur}}, u) + d(u, v_{\text{vor}(u)})$
21:             **end if**
22:         **end if**
23:     **end for**
24: **end while**

In the analysis to follow (in Section 5) we move back and forth between a graph and its dual. For this we need the following definitions.

Definition 3. *Given a path* $P = (u_0, u_1, \ldots, u_h)$, *the* Voronoi path *of* P *is the sequence of vertices* $P^* = (v_{\text{vor}(u_0)}, v_{\text{vor}(u_1)}, \ldots, v_{\text{vor}(u_h)})$.

Note that the Voronoi path P^* may not necessarily be simple, as multiple consecutive occurrences of nodes $v_{\text{vor}(u_i)}$ are possible in P^*. They are treated as a single occurrence, and such paths are deemed to be equivalent.

Lemma 1. *For any path* $P = (u_0, \ldots, u_h)$ *in an undirected graph* $G = (V, E, \omega)$, *the corresponding Voronoi path* P^* *exists and is unique.*

Proof. Suppose that there is no such path P^* in G^*. This implies that there exist pairs of nodes u_i, u_{i+1} on the path P for which $v_{\text{vor}(u_i)} \neq v_{\text{vor}(u_{i+1})}$ and $(v_{\text{vor}(u_i)}, v_{\text{vor}(u_{i+1})}) \notin E^*$. As u_i, u_{i+1} are consecutive nodes on the path P, we know that $(u_i, u_{i+1}) \in E$. This contradicts the definition of the Voronoi dual (Def. 2), since $(u_i, u_{i+1}) \in E$ and $v_{\text{vor}(u_i)} \neq v_{\text{vor}(u_{i+1})}$ together imply that $(v_{\text{vor}(u_i)}, v_{\text{vor}(u_{i+1})}) \in E^*$. P^* is unique since each node u_i on the path belongs to exactly one Voronoi region, corresponding to exactly one Voronoi node $v_{\text{vor}(u_i)}$.

Definition 4. *For a path* P^* *in the Voronoi dual* G^* *of a graph* G, *the* Voronoi sleeve *is the subgraph of* G *induced by the nodes in the union of all Voronoi regions* V_i *for which its Voronoi node* v_i *lies on* P^*,

$$\mathsf{Sleeve}_{(G,G^*)}(P^*) := G\left[\bigcup_{v_i \in P^*} V_i\right].$$

With the definitions at hand we can now state the approximation method.

3   The Method

In *preprocessing*, each node is selected as a Voronoi site independently at random with probability p, and the Voronoi dual is computed for the selected sites (Algorithm 2). For the sake of exposition, we treat the computation of the Voronoi dual as a 'black box', denoted by `ComputeVoronoiDual`.

Algorithm 2. Preprocessing

Input: graph $G = (V, E, \omega)$, sampling rate $p \in [0, 1]$.
Output: Voronoi dual G^* with Voronoi nodes selected independently at random with probability p.

1: Random sampling: Generate the set of Voronoi nodes by selecting each node of V independently at random: $\forall v \in V, \Pr[v \in K] = p$.
2: Compute a Voronoi dual $G^* = (K, E^*, \omega^*)$ using the modified version of Erwig's algorithm [Erw00, Section 3.1] as shown in Algorithm 1.
   $G^* := \texttt{ComputeVoronoiDual}(G, K)$
3: Return G^*.

Lemma 2. *For a graph $G = (V, E)$ with $n := |V|$ and $m := |E|$, Algorithm 2 takes time proportional to that of Dijkstra's single source shortest path algorithm.*

Proof. Erwig's variant of Dijkstra's algorithm computes the graph Voronoi diagram in a worst-case time proportional to Dijkstra's algorithm [Erw00, Theorem 2]. The only modification of Algorithm 1 compared to Erwig's variant is the following: for each node, at the time it is settled, all its neighbors are inspected. Therefore, each edge is additionally considered two times in total. This yields the same asymptotic running time.

The preprocessing time complexity is proportional to the cost of computing one single source shortest path tree. Details are discussed in Section 4.

At *query* time, given a graph G and its Voronoi dual G^* we answer (approximate) shortest path queries between source s and target t, by first searching for a shortest path $SP_{G^*}(v_{\mathsf{vor}(s)}, v_{\mathsf{vor}(t)})$ in the smaller Voronoi dual G^*. This path determines the subgraph $\mathcal{S} = \mathsf{Sleeve}(SP_{G^*}(v_{\mathsf{vor}(s)}, v_{\mathsf{vor}(t)}))$, whose shortest path $SP_{\mathcal{S}}(s, t)$ approximates the shortest path $SP_G(s, t)$ in G. The shortest path in the Voronoi dual guides the Dijkstra search in the original graph. For a pseudo-code description, see Algorithm 3.

Algorithm 3. Query

Input: Graph G, Voronoi dual G^*, Source s, Target t.
Output: an approximate shortest path P from s to t.

1: Find Voronoi source $v_{\mathsf{vor}(s)}$ from s and Voronoi target $v_{\mathsf{vor}(t)}$ from t. If thereby a shortest path $SP_G(s,t)$ has been found, return it.
2: Compute a shortest path from $v_{\mathsf{vor}(s)}$ to $v_{\mathsf{vor}(t)}$ in the Voronoi dual G^*: $SP_{G^*}(v_{\mathsf{vor}(s)}, v_{\mathsf{vor}(t)})$.
3: Compute the Voronoi sleeve

$$\mathcal{S} := \mathsf{Sleeve}(SP_{G^*}(v_{\mathsf{vor}(s)}, v_{\mathsf{vor}(t)})).$$

4: Compute a shortest path from s to t in the Voronoi sleeve, $SP_{\mathcal{S}}(s,t)$.
5: Return $P = SP_{\mathcal{S}}(s,t)$.

The running time of Algorithm 3 depends on G and p. Let N^* and M^* denote the random variables measuring the number of nodes and edges of the Voronoi dual. Clearly $E[N^*] = p \cdot n$. The expected query time *without* refinement (computing the shortest path in the Voronoi sleeve) is at most $O(N^* \lg N^* + M^*)$. The time for the refinement step depends on the size of the Voronoi sleeve. The analysis will show that the refinement step is not necessary for the approximation ratio to hold for long distance queries; however, it makes a practical difference for the quality of paths. For $p = O(n^{-2/3})$, $E[N^*] = O(n^{1/3})$, and thus we can afford to compute all-pairs shortest path distances in the Voronoi dual G^* in overall linear expected time. This allows for constant-time approximate distance queries.

4   Computational Complexity

In this section we study the cost of computing a Voronoi dual. Recall that in Erwig's algorithm [Erw00, Section 3.1] the graph Voronoi diagram is constructed with a single Dijkstra search. A heap is used to store the shortest path distances from nodes to their closest Voronoi node. Conceptually, a dummy node with a zero-weighted edge to each of the Voronoi nodes is added, the dummy node is inserted into the heap, and the Dijkstra single source shortest path search is executed. The running times of different implementations of Dijkstra's algorithm depend on the priority queue employed (see Table 1). Using Fibonacci heaps [FT87], Dijkstra's algorithm takes time $O(m + n \lg n)$.

Erwig also claims a time lower bound of $\Omega(\max(n, (n-k) \lg k))$ [Erw00, Theorem 1]. The lower bound simplifies to $\Omega(n \lg n)$ when the number of Voronoi nodes is assumed to be $k = n^C$ for a fixed choice of $C \in (0,1)$. Assuming that all edges must be inspected at construction time, this lower bound would be tight. The bound is information theoretic: for a connected graph, each node $w \in V \setminus K$ is in exactly one of the k regions V_i. Encoding one instance out of these k^{n-k} possibilities requires $\lg k^{n-k} = (n-k) \lg k$ bits.

Table 1. Running times for different implementations of Dijkstra's algorithm, excerpted from [Tho99, p. 364]. The algorithms in the first two rows work for both the pointer machine and the RAM model. The analysis of the algorithms from row 3 onwards only works in the RAM model.

Time	Reference
$O(m \lg n)$	[Wil64]
$O(m + n \lg n)$	[FT87]
$O(m \sqrt{\lg n})$	[FW93]
$O(m + n \frac{\lg n}{\lg \lg n})$	[FW93]
$O(m \lg \lg n)$	[Tho00b]
$O(m + n \lg^{1/2+\epsilon} n)$	[Tho00b]
$O(m + n \sqrt{\lg n \lg \lg n})$	[Ram96]
$O(m + n \lg^{1/3+\epsilon} n)$	[Ram97]
$O(m + n \lg \lg n)$	[Tho04b]

For some graphs with special properties, Erwig's lower bound may not apply. Eppstein and Goodrich [EG08] presented a linear-time algorithm to compute the Voronoi diagram for road networks satisfying certain geometric properties. Also, the lower bound may not hold under different models of computation, such as the word RAM model. This model assumes that basic operations such as adding two words requires a single time step, and that the time compexity is the number of word operations executed. The space complexity is the number of words of storage required, assuming that any identifier (such as a node label) or value (such as a distance) can be contained in a single word. Under the word RAM model, the implementation of Dijkstra's algorithm by Thorup [Tho04b] requires only $O(m + n \lg \lg n)$-time.

Corollary 1. *The graph Voronoi dual can be computed in time $O(m + n \lg \lg n)$ in the word RAM model.*

Note that the time upper bound under the word RAM model does not contradict Erwig's information-theoretic lower bound [Erw00, Theorem 1] of $\Omega(n \lg n)$ bits.

Computing a graph Voronoi dual does not actually require the use of Dijkstra's algorithm — any single source shortest path algorithm (including parallel and distributed algorithms) can be used to compute a graph Voronoi dual as follows. Instead of an adapted Dijkstra search, we may also

1. augment G by introducing a dummy node v_d connected to each of the Voronoi nodes with an edge of length zero,
2. run any single source shortest path algorithm in the augmented graph G' with v_d as its source, and
3. explore the search tree rooted at v_d by following shortest path edges only.

This last step simulates a Dijkstra search by following the single source shortest path tree without using any expensive decrease-key operations (these operations have to be avoided to reduce the worst-case running time [Tho00b, Tho07]); a First-In-First-Out queue with constant time for the enqueue and dequeue operations is sufficient. For a pseudo-code description, see Algorithm 4. Although the construction is mainly of theoretical interest, it may be useful for example for parallel or distributed algorithms and for software that must rely on certain libraries.

Algorithm 4. $\mathtt{ComputeVoronoiDual}(G = (V, E), K \subseteq V)$

1: Let $G' := (V', E')$ with $V' = V \cup \{v_d\}$ and $E' = E \cup \{(v_d, v) : v \in K\}$ with $\omega'(v_d, v) = \delta$ (one would set $\delta = 0$ if possible; if only positive edge are allowed, other values work as well)

2: $\mathcal{D} := \mathtt{SSSP}(G', v_d)$, where $\mathcal{D}$ is the distance vector storing the distance from v_d to each node $u \in V'$

3: **for** $i := 1$ **to** $k = |K|$ **do**

4:     $\mathrm{vor}(v_i) := i$

5:     $\mathtt{FIFO.enqueue}(v_i)$

6: **end for**

7: **while** $\neg\mathtt{FIFO.empty}$ **do**

8:     $u_{\mathrm{cur}} := \mathtt{FIFO.dequeue}$

9:     **for** $u \in \Gamma(u_{\mathrm{cur}})$ **do**

10:         **if** $\mathcal{D}(u) = \mathcal{D}(u_{\mathrm{cur}}) + \omega(u, u_{\mathrm{cur}})$ **and** $\mathrm{vor}(u) = \mathrm{undef}$ **then**

11:             $\mathrm{vor}(u) := \mathrm{vor}(u_{\mathrm{cur}})$

12:             $\mathtt{FIFO.enqueue}(u)$

13:         **else if** $\mathrm{vor}(u) \neq \mathrm{undef}$ **and** $\mathrm{vor}(u) \neq \mathrm{vor}(u_{\mathrm{cur}})$ **then**

14:             **if** $(v_{\mathrm{vor}(u_{\mathrm{cur}})}, v_{\mathrm{vor}(u)}) \notin E^*$ **then**

15:                 $E^* := E^* \cup \{(v_{\mathrm{vor}(u_{\mathrm{cur}})}, v_{\mathrm{vor}(u)})\}$

16:                 $\omega^*(v_{\mathrm{vor}(u_{\mathrm{cur}})}, v_{\mathrm{vor}(u)}) := \infty$

17:             **end if**

18:             **if** $\omega^*(v_{\mathrm{vor}(u_{\mathrm{cur}})}, v_{\mathrm{vor}(u)}) > \mathcal{D}(v_d, u_{\mathrm{cur}}) - \delta + \omega(u_{\mathrm{cur}}, u) + \mathcal{D}(u, v_d) - \delta$ **then**

19:                 $\omega^*(v_{\mathrm{vor}(u_{\mathrm{cur}})}, v_{\mathrm{vor}(u)}) := \mathcal{D}(v_d, u_{\mathrm{cur}}) - \delta + \omega(u_{\mathrm{cur}}, u) + \mathcal{D}(u, v_d) - \delta$

20:             **end if**

21:         **end if**

22:     **end for**

23: **end while**

Note that, if a single source shortest path algorithm $\mathcal{A}$ works for a special class of graphs $\mathcal{G}$, the augmented graph G' may not necessarily be in $\mathcal{G}$, and thus algorithm $\mathcal{A}$ cannot be used in general. For example, for planar graphs, the $O(n)$-time algorithm by Henzinger et al. [HKRS97] cannot be applied directly to compute the Voronoi diagram since planarity may be violated by adding a dummy node. In the particular case of the algorithm by Henzinger et al., however, the analysis of the running time depends on separators which do admit the introduction of a dummy node.

Theorem 1. *Using any general single source shortest path algorithm with running time $t(n, m)$, Algorithm 4 computes a graph Voronoi dual in time $O(n + m + t(n, m))$.*

Proof. After running the SSSP algorithm in time $t(n, m)$, Algorithm 4 visits every node exactly once and every edge exactly twice (once for each end point).

For undirected graphs we may use the $O(m)$-time SSSP algorithm by Thorup [Tho99, Tho00a].

Corollary 2. *For undirected graphs, the graph Voronoi dual can be computed in time $O(m + n)$ in the word RAM model.*

Corollary 3. *For a graph $G = (V, E)$ with $n := |V|$ and $m := |E|$, Algorithm 2 takes time proportional to that of Dijkstra's single source shortest path algorithm.*

5   Stretch Analysis

In this section, we prove that the expected path length approximation ratio is logarithmic in the number of edges of an exact shortest path.

Theorem 2. *For shortest paths having h edges, Algorithm 3, given a graph and its Voronoi dual with sampling rate p (constructed by Algorithm 2), has expected approximation ratio $O(\lg_{1/(1-p)} h)$.*

The path $SP_S(s, t)$ found by the algorithm is an approximation, since it is possible that no actual shortest path $SP_G(s, t)$ lies entirely within the Voronoi sleeve S. We explain how this is possible, and give an upper bound on the expected length $\ell(SP_S(s, t))$. For this purpose, we prove relationships between the lengths of simple paths P and their corresponding Voronoi paths P^*. The stretch of a path P^* depends on the number and distribution of Voronoi nodes on the path P. In particular, the stretch depends linearly on the largest interval between two Voronoi nodes on the path.

Definition 5. *For a path $P = (u_0, u_1, \ldots, u_h)$ in a graph $G = (V, E, \omega)$, and a set of Voronoi nodes $K \subseteq V$, two Voronoi nodes v_i, v_j on P are called* consecutive *if the subpath between v_i and v_j does not contain another Voronoi node. The* gap *g between two consecutive Voronoi nodes on the path is defined as the number of edges of this subpath. The* largest gap *of a path is the maximum over all gaps between two consecutive Voronoi nodes on the path.*

To simplify the analysis, we initially assume that s and t are Voronoi nodes. Later, we will relax this restriction.

We wish to prove that the stretch is at most the size of the largest gap $\bar{h}$ between two Voronoi nodes on the path $SP_G(s, t)$. For the analysis we fix a shortest path $SP_G(s, t) = (s, u_1, u_2, \ldots, u_{h-1}, t)$. If the corresponding Voronoi path $(SP_G(s, t))^*$ is a shortest path from s to t in the Voronoi dual, then the Voronoi sleeve S also contains $SP_G(s, t)$. Figure 2 gives an example for which $(SP_G(s, t))^*$ is not a shortest path in the dual.

In Lemma 3, for any simple path P, we give a worst-case bound on the length of the corresponding Voronoi path. P^* can have maximal stretch if there is no Voronoi node among the intermediate nodes and the corresponding Voronoi nodes have maximal distance (while still satisfying the Voronoi condition).

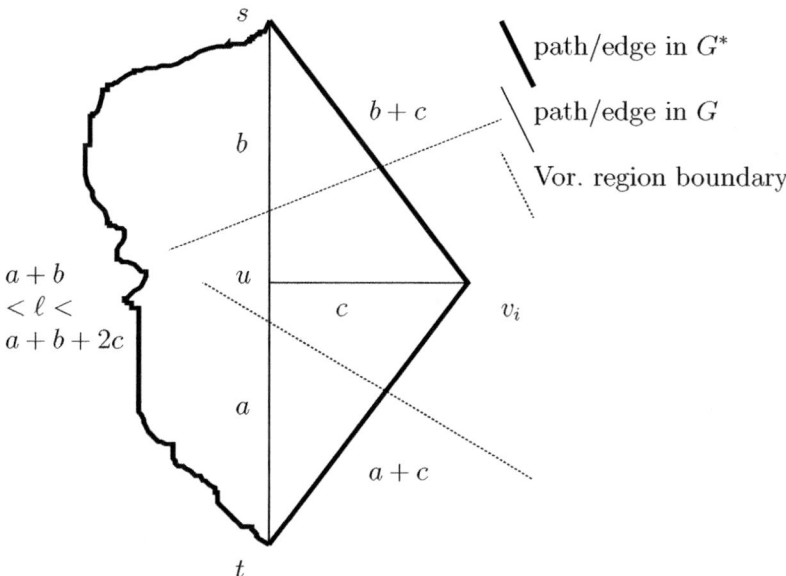

Fig. 2. s, t, and v_i are Voronoi nodes. The shortest path from s to t leads through u, which is in v_i's Voronoi region (if $c < a$ and $c < b$), and paths in the Voronoi dual pass through v_i. If $\ell < a + b + 2c$, the shortest path in the Voronoi dual SP_{G^*} takes the left-hand route, and the Voronoi sleeve $\mathcal{S}$ does not contain u.

Lemma 3. *Given a simple path $P = (s, u_1, \ldots, u_{h-1}, t)$ between two Voronoi nodes $s = u_0$ and $t = u_h$ with h edges and length $\ell(P)$, the corresponding Voronoi path P^* in the Voronoi dual G^* has at most length $\ell(P^*) \leq h \cdot \ell(P)$. This upper bound is tight.*

Proof. The path contains $h - 1$ intermediate nodes and h edges and therefore passes through at most $h + 1$ different Voronoi regions. Out of these, at most $h - 1$ regions are 'interfering' regions, meaning that the original shortest path does not lead through the corresponding Voronoi nodes but the shortest Voronoi path does. The path length $\ell(P)$ in the original graph is the sum of the edge weights $\ell(P) := d(s, t) = \sum_{k=0}^{h-1} \omega(u_k, u_{k+1})$. The length $d^*(v_{\mathsf{vor}(u_k)}, v_{\mathsf{vor}(u_{k+1})})$ of an edge between two Voronoi nodes on the path P^* can be bounded as follows (see Figure 3):

$$d^*(v_{\mathsf{vor}(u_k)}, v_{\mathsf{vor}(u_{k+1})}) \leq d(v_{\mathsf{vor}(u_k)}, u_k) + \omega(u_k, u_{k+1}) + d(u_{k+1}, v_{\mathsf{vor}(u_{k+1})})$$

From the Voronoi condition, we observe that $\forall j : d(u_k, v_{\mathsf{vor}(u_k)}) \leq d(u_k, v_{\mathsf{vor}(u_j)})$. Due to the assumption that s and t are also Voronoi nodes, this also holds for source and target. That is,

$$d(u_k, v_{\mathsf{vor}(u_k)}) \leq d(s, u_k)$$
$$d(u_k, v_{\mathsf{vor}(u_k)}) \leq d(u_k, t)$$
$$= d(v_{\mathsf{vor}(u_k)}, u_k)$$

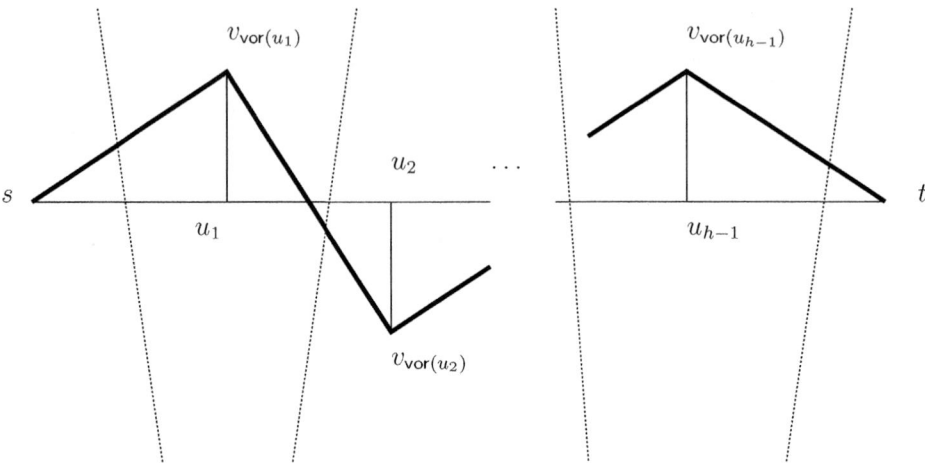

Fig. 3. The shortest path between two Voronoi nodes s and t with $h-1$ intermediate nodes $u_1, \ldots, u_{h-1}$. The distance between two Voronoi nodes that are adjacent in the Voronoi dual is at most $\omega^*(v_{\text{vor}(u_k)}, v_{\text{vor}(u_{k+1})}) \leq d(v_{\text{vor}(u_k)}, u_k) + \omega(u_k, u_{k+1}) + d(u_{k+1}, v_{\text{vor}(u_{k+1})})$.

This yields:

$$
\begin{aligned}
\ell(P^*) \leq d^*(s,t) = {}& d^*(s, v_{\text{vor}(u_1)}) \\
&+ \sum_{k=1}^{h-2} \left[d(v_{\text{vor}(u_k)}, u_k) + \omega(u_k, u_{k+1}) + d(u_{k+1}, v_{\text{vor}(u_{k+1})}) \right] \\
&+ d^*(v_{\text{vor}(u_{h-1})}, t) \\
\leq {}& \omega(s, u_1) + d(u_1, v_{\text{vor}(u_1)}) \\
&+ \sum_{k=1}^{h-2} \left[d(v_{\text{vor}(u_k)}, u_k) + d(u_{k+1}, v_{\text{vor}(u_{k+1})}) \right] \\
&+ \sum_{k=1}^{h-2} \omega(u_k, u_{k+1}) \\
&+ d(v_{\text{vor}(u_{h-1})}, u_{h-1}) + \omega(u_{h-1}, t) \\
\leq {}& d(s,t) + \sum_{k=1}^{h-1} \left[d(s, u_k) + d(u_k, t) \right] \\
= {}& h \cdot \ell(P)
\end{aligned}
$$

There exist constructions for which the bound can be shown to be tight. For example, for any choice of $a > \epsilon > 0$, the edge weights of G may be chosen such that $d(u_k, v_{\text{vor}(u_k)}) = a - \epsilon$, $\omega(u_k, u_{k+1}) = \epsilon$, and $\omega(s, u_1) = \omega(u_{h-1}, t) = a$. Path P has length $2a + (h-2)\epsilon$, and the Voronoi path P^* has length $2a + (h-2)\epsilon + 2(h-1) \cdot (a - \epsilon)$. As $\epsilon \to 0$, the ratio $\ell(P^*)/\ell(P) \to h$.

If in addition to the endpoints there are Voronoi nodes on the shortest path, the maximum stretch is guaranteed to be smaller than the number of edges on the shortest path. In the following lemma, we prove that the maximum stretch is proportional to the largest gap between Voronoi nodes on the path. The proof is a simple composition of Lemma 3, and is supported by the illustration in Figure 3.

Lemma 4. *Let $P = (v_i, u_1, \ldots, u_{h-1}, v_j)$ be a simple path of length $\ell(P)$ between two Voronoi nodes $v_i = u_0$ and $v_j = u_h$. Let $\bar{h}$ denote the largest gap of P. The corresponding Voronoi path P^* in the Voronoi dual G^* has at most length $\ell(P^*) \leq \bar{h} \cdot \ell(P)$. This upper bound is tight.*

Proof. Suppose there are $2 + \nu$ Voronoi nodes $u_k = v_{\mathrm{vor}(u_k)}$ on the path. The remaining $h - 1 - \nu$ nodes are non-Voronoi nodes. We cut the path P into subpaths P_k between Voronoi nodes. Let h_k denote the number of edges between two consecutive Voronoi nodes, which is the number of edges of P_k. The Voronoi path is composed of $1 + \nu$ segments P_k between Voronoi nodes ($\sum_{k=0}^{\nu} \ell(P_k) = P$, $\sum_{k=0}^{\nu} h_k = h, \forall k : h_k \leq \bar{h}$). Composition of Lemma 3 leads to the following bound on the path length:

$$\sum_{k=0}^{\nu} h_k \ell(P_k) \leq \sum_{k=0}^{\nu} \max_{\kappa \in \{0,\ldots,\nu\}} h_\kappa \ell(P_k) \leq \bar{h} \cdot \ell(P).$$

Tightness can be shown with the same example as in the proof of Lemma 3.

Lemma 6 gives an upper bound on the expected size of the largest gap. We use the following lemma by Szpankowski and Rego [SR90] concerning the maximum of geometric random variables.

Lemma 5 (Szpankowski and Rego [SR90, eq. (2.6) and (2.12)]). *Let X_i, $i = 1, 2, \ldots, n$ be a set of i.i.d. random variables distributed according to the geometric distribution with parameter p. That is, for every $i = 1, 2, \ldots, n$ and $k \in \mathbb{N}^+$,*

$$\Pr[X_i = k] = (1 - p)^{k-1} p$$
$$\mathsf{E}[X_i] = p^{-1}$$
$$\mathsf{E}[X_i^2] = (2 - p)p^{-2}.$$

Let $M_n = \max\{X_1, X_2, \ldots, X_n\}$. The expected value of M_n is

$$\mathsf{E}[M_n] = -\sum_{k=1}^{n} (-1)^k \binom{n}{k} \frac{1}{1 - (1-p)^k}$$
$$= \lg_{1/(1-p)} n + O(1).$$

Lemma 6. *In a path of length $h - 1$, where each node has been selected as a Voronoi node independently at random with probability p, the longest sequence of non-Voronoi nodes is of expected length at most $O(\lg_{1/(1-p)} h)$.*

Proof. The path can be seen as a sequence of coin tosses, for which we want to bound the expected length of the longest sequence of tails. This problem is known as the Longest Success-Run [EMK97, Ch. 8.5]. We wish to bound the expectation of the maximum of N independent geometric random variables with probability p and sum $h - 1 - N$ (N itself being a random variable).

To derive a bound on the expectation, we observe that by dropping the sum condition, and by taking the maximum over $h \geq N$ random variables, the maximum value obtained can only increase.

As of Lemma 5, the expectation of the maximum of h geometric random variables with probability p is known to be at most $O(\lg_{1/(1-p)} h)$.

We now combine Lemmas 3, 4, and 6 to prove Theorem 2.

Proof (Proof of Theorem 2). Consider first the case where s and t are both Voronoi nodes.

Let $\bar{h}$ denote the largest gap of some shortest path $SP_G(s,t)$. Lemma 4 implies that the corresponding Voronoi path $(SP_G(s,t))^*$ has length at most $\bar{h} \cdot \ell(SP_G(s,t))$. Trivially, the shortest path in the Voronoi dual is of length no more than that of the Voronoi path; that is, $\ell((SP_G(s,t))^*) \geq \ell(SP_{G^*}(s,t))$. The path $SP_{G^*}(s,t)$ in the Voronoi dual corresponds to a path P' of the same length in the Voronoi sleeve $\mathsf{Sleeve}(SP_{G^*}(s,t))$. Therefore,

$$\ell(SP_S(s,t)) \leq \ell(P')$$
$$= \ell(SP_{G^*}(s,t))$$
$$\leq \ell((SP_G(s,t))^*)$$
$$\leq \bar{h} \cdot \ell(SP_G(s,t)).$$

Recall that nodes are independently selected as Voronoi nodes with sampling rate p. For a shortest path with h edges, the expected largest gap $\bar{h}$ is at most $O(\lg_{1/(1-p)} h)$ by Lemma 6.

For the case where either s or t (or both) are not Voronoi nodes, if the path returned by Algorithm 3 has been found in Step 1, it is optimal, and the result holds trivially. For the remainder of the proof we assume that the shortest path has not been found in Step 1. In this case, the path returned is at most as long as the shortest path P_{vor} in G from s to t having $SP_{\mathsf{Sleeve}(SP_{G^*}(v_{\mathsf{vor}(s)}, v_{\mathsf{vor}(t)}))}(v_{\mathsf{vor}(s)}, v_{\mathsf{vor}(t)})$ as a subpath. In the following, we derive an upper bound on $\ell(P_{\mathsf{vor}})$ with respect to the number of edges on the shortest path between s and t, denoted by h'. We have that

$$\ell(P_{\mathsf{vor}}) \leq d(s, v_{\mathsf{vor}(s)}) + d^*(v_{\mathsf{vor}(s)}, v_{\mathsf{vor}(t)}) + d(v_{\mathsf{vor}(t)}, t).$$

Since the shortest path from s to t has not already been found directly in Step 1, it must be true that both $d(s, v_{\mathsf{vor}(s)}) \leq d(s,t)$ and $d(s, v_{\mathsf{vor}(s)}) \leq d(s,t)$. It remains to bound the distance between $v_{\mathsf{vor}(s)}$ and $v_{\mathsf{vor}(t)}$ in the dual graph.

Observe that augmenting the graph G with one edge $(u, v_{\mathsf{vor}(u)})$ of weight $d(u, v_{\mathsf{vor}(u)})$ for each non-Voronoi node $u \in V \setminus K$ affects neither the Voronoi diagram nor the Voronoi dual, since the nodes on the shortest path from $v_{\mathsf{vor}(u)}$ to u cannot be interfered with by another Voronoi node.

In the augmented primal graph, by the triangle inequality, we have that $d(v_{\mathsf{vor}(s)}, v_{\mathsf{vor}(t)}) \leq d(v_{\mathsf{vor}(s)}, s) + d(s, t) + d(t, v_{\mathsf{vor}(t)}) \leq 3d(s, t)$ using a path with at most $1 + h' + 1$ edges. Therefore, the expected distance $d^*(v_{\mathsf{vor}(s)}, v_{\mathsf{vor}(t)})$ is also bounded by $O(\lg h') \cdot 3d(s, t)$. The bound for P_{vor} follows directly.

This concludes the proof of Theorem 2.

6   Experiments

In the following, we provide an experimental evaluation for our implementation of the Voronoi shortest path approximation method. The preprocessing and query times are compared with those of Dijkstra's algorithm and with those of related but exact methods.

6.1   Algorithms

Benchmarking. As the methods in our study were developed and compiled on different computers and architectures, a direct comparison with reported query times would not be meaningful. We measure the performance of the methods against the bidirectional version of Dijkstra's algorithm, in terms of the ratio of the number of nodes settled by Dijkstra's algorithm over the number of nodes settled by the Voronoi method. This ratio, which we will refer to as the *speed-up* of the method, can be used to evaluate the performance of Steps 1, 2, and 4 of Algorithm 3. In addition, we count the number of marked regions to account for Step 3.

The use of the Voronoi sleeve in Steps 3 and 4 of Algorithm 3 leads to practical improvements in accuracy; however, the example in Figure 2 shows that for general graphs the worst-case stretch does not improve. For all the experiments, we evaluate the method once using the refinement step and once with these Voronoi sleeve steps omitted. For the second type of queries, the reported distance is the sum of the distances from the query source to the Voronoi source, from the Voronoi source to the Voronoi target, and from the Voronoi target to the query target, as computed in Steps 1 and 2 of Algorithm 3.

Voronoi method. Our method using the Voronoi dual can be parameterized using the sampling probability p, the value of which determines the trade-off between approximation quality and speed-up. For the evaluation, we consider three values of the sampling probability — $p = 1/2$, $p = n^{-1/2}$, and $p = n^{-2/3}$ — that produce Voronoi nodesets of expected sizes $n/2$, $\sqrt{n}$, and $\sqrt[3]{n}$ respectively. The variants are referred to as VORHALF, VORROOT, and VORCUBERT.

Other methods. Sanders and Schultes [SS07a, Table 1] provide a detailed overview of methods for accelerated point-to-point shortest path queries in road networks. Bauer et al. [BDS+08, p. 13] list another set of methods and compare their performance on several transportation networks. We select some of the fastest methods for comparison with our algorithm. Unless stated otherwise, we will use the naming conventions of [SS07a, BDS⁺08] to refer to these methods.

- Highway Hierarchies (HH) [SS06] are based on the observation that a certain class of edges (the 'highway' edges) tend to have greater representation among the portion of the shortest paths that are not in the vicinity of either the source or target. A recursive computation of these edges, paired with a contraction step, leads to a hierarchy of graphs that enables an impressive speed-up at query time. HH+dist denotes a variant of HH where all higher levels with at most $O(\sqrt{n})$ nodes are replaced by a single distance table. HH+dist+A* is HH combined with A* search and implemented with distance tables [DSSW06]. Highway Node Routing (HNR) [SS07b] is another variant of the Highway Hierarchies strategy.
- In the same spirit as HH, Transit Node Routing (TNR) [BFM⁺07] identifies a set of nodes (called 'transit' nodes) that often occur on shortest paths. A table storing the distances between all pairs of these nodes allows any shortest path distance to be computed with a small number of table lookups. Two variants are listed: TNR-eco with economical space consumption, and TNR-gen with generous space consumption.
- The Arc-Flag method [Lau04] computes a partition of the graph and then, for each component and for each shortest path ending in that component, it labels the first edge. A variant of this method, SHARC [BD08], incorporates techniques developed for Highway Hierarchies.
- Contraction Hierarchies (CHHNR) [GSSD08] is an extension of highway hierarchies in which the graph is further simplified using contraction operations. Many variants have been proposed; we consider only the variant with the fastest preprocessing time, CHHNR_EDS1235, and the variant with the best speed-up, CHHNR_EVSQWL. The CHASE method [BDS⁺08] integrates the Contraction Hierarchies and Arc-Flag methods.
- A method based on A* search by Goldberg and Harrelson [GH05], which we will refer to as simply A*, is one of the first methods with reasonable preprocessing time and good speed-up.
- ALT-m16 [DW07] is a variant of ALT [GW05], which in turn is a combination of A*, Landmarks, and speed-up techniques based on the triangle inequality. CALT-m16 and CALT-a64 [BDS⁺08] are two variants of a method that combines ALT and Contraction Hierarchies.

6.2   Data Sets

For the sake of comparison, we consider transportation networks that were used by Sanders and Schultes [SS07a] and Bauer et al. [BDS⁺08, BD08] in their evaluations. In addition, to demonstrate that our method is effective for more general graphs, we run experiments with a social network, a citation graph, a router network, and protein interaction networks as data sets. The node degrees of these graphs seem to follow a power-law distribution [Mit03].

Road networks. The road network of Western Europe has been made available for scientific use by the company PTV AG. It covers 14 countries and, with its massive size of 18,010,173 nodes and 42,560,279 directed edges, it serves as an important benchmark for shortest path queries. In order to apply the Voronoi method, we convert the graph into an undirected form. There are two different edge weightings, one representing geographical distances and the other representing driving time. We conduct experiments for both.

Public transportation. We also conduct experiments for three European public transportation networks:

1. long railway connections in Europe, with 1,586,862 nodes and 2,402,352 directed edges,
2. the bus network of the Rhein-Main-Verkehrsverbund RMV, with 2,278,066 nodes and 3,417,084 directed edges, and
3. the bus network of the Verkehrsverbund Berlin Brandenburg VBB, with 2,600,818 nodes and 3,901,212 directed edges.

The graphs considered by [BDS$^+$08, BD08] differ slightly from those used for experimentation with the Voronoi method.

The numbers of nodes and edges of the RMV and VBB input graphs are nearly identical; however, the long railway graph used in our experimentation has 33% more nodes and edges than in [BDS$^+$08, BD08]. Again, for the Voronoi experimentation, the graphs were converted into an undirected form.

Social networks. We extracted the DBLP computer science bibliography [Ley02] co-author graph from an official XML version downloaded on 24 August 2008. In the graph, two authors are connected by an edge if they have at least one joint publication. This yielded an undirected graph, from which we selected the largest connected component. The final graph is unweighted and consists of 511,163 nodes and 1,871,070 edges.

Router topology. CAIDA maintains data on the router-level topology of a portion of the Internet [Coo03]. After cleaning we obtained an undirected, unweighted graph with 190,914 nodes and 607,610 edges.

Citation graph. The citations for 27,400 publications in the high energy physics research literature were used as a data set in the KDD Cup 2003 competition [GSDF03]. From these citations, we constructed an undirected, unweighted graph with 352,542 edges.

Protein interactions. The Database of Interacting Proteins [SMS$^+$02] catalogs experimentally determined interactions among proteins. We extracted the largest connected component, consisting of 19,928 nodes and 82,406 edges. BioGRID is a general repository for interaction data sets [SBR$^+$06] from which we extracted the largest connected component, consisting of 4,039 nodes and 43,854 edges.

6.3   Experimental Setting

In this section we describe the experimental setting for the Voronoi method. The implementation is written in C++ and executed on one core of a 2x2.66 GHz Dual-Core Intel Xeon Desktop with 6 GB 800 MHz DDR2 FB-DIMM running Mac OS X 10.5.6.

Table 2. Experimental results for the Voronoi method

method	preprocessing [s]	without sleeve		with sleeve	
		speed-up	stretch	speed-up	stretch
PTV European road network, driving time, 18,010,173 nodes, 42,560,279 edges					
VorHalf	31.7686±4.4436	2.6061± 0.0734	1.0394±0.0131	2.5878± 0.0750	1.0111±0.0062
VorRoot	40.5296±3.6423	3,518.0645± 725.2776	1.6613±0.2078	4.9991± 4.9017	1.1291±0.0783
VorCubeRt	31.3372±2.8181	39,918.4988±14,207.5395	1.5544±0.4292	1.5863± 1.1123	1.0405±0.0597
PTV European road network, geographical distance, 18,010,173 nodes, 42,560,279 edges					
VorHalf	29.8365±4.3576	2.6266± 0.0558	1.0307±0.0095	2.5800± 0.0627	1.0139±0.0057
VorRoot	34.2785±3.0609	3,672.4070± 511.1418	1.1821±0.0960	5.9212± 7.9921	1.0390±0.0249
VorCubeRt	22.5531±2.0284	42,266.6442±13,530.5983	1.2882±0.5384	1.6383± 1.4232	1.0141±0.0291
Public transportation, long distance railway, 1,586,862 nodes, 2,402,352 edges					
VorHalf	2.0499±0.1998	1.9511± 0.1231	1.0180±0.0227	1.8972± 0.1367	1.0080±0.0143
VorRoot	1.9086±0.0946	363.8390± 153.4644	1.3813±0.2848	2.8527± 3.3113	1.0829±0.0971
VorCubeRt	1.7633±0.0860	2,116.0373± 1,251.1773	1.5167±0.6610	1.2599± 0.5990	1.0247±0.0658
Public transportation, RMV, 2,278,066 nodes, 3,417,084 edges					
VorHalf	3.7714±0.4064	1.9892± 0.1813	1.0290±0.0255	1.9315± 0.1766	1.0104±0.0131
VorRoot	3.7455±0.2158	789.2912± 328.2714	1.2972±0.2591	3.1802± 5.4237	1.0644±0.0864
VorCubeRt	3.4120±0.1633	5,973.7950± 3,748.1389	1.3522±0.6003	1.3089± 0.9703	1.0204±0.0583
Public transportation, VBB, 2,600,818 nodes, 3,901,212 edges					
VorHalf	4.1409±0.4180	1.9881± 0.6476	1.0335±0.0248	1.9313± 0.5172	1.0075±0.0097
VorRoot	4.0242±0.2914	866.8917± 405.4821	1.4042±0.2516	3.7864± 7.6010	1.0834±0.1000
VorCubeRt	3.7145±0.2333	7,373.2971± 4,742.2783	1.4375±3.3690	1.3427± 1.2759	1.0244±0.0660
DBLP co-authorship, 511,163 nodes, 1,871,070 edges					
VorHalf	0.9145±0.0431	1.3576± 1.4690	1.2093±0.1805	1.3447± 1.4364	1.1419±0.1468
VorRoot	0.8376±0.0430	37.7082± 53.2992	1.9323±0.3591	11.4432±14.8387	1.3954±0.2850
VorCubeRt	0.6041±0.0312	143.8757± 208.7946	2.0033±0.3630	9.9616±12.5412	1.2881±0.2406
CAIDA router topology, 190,914 nodes, 607,610 edges					
VorHalf	0.3050±0.0154	1.3164± 1.1720	1.1810±0.1703	1.2972± 1.1074	1.1283±0.1359
VorRoot	0.1793±0.0092	42.4832± 54.6527	1.7845±0.3533	7.8865± 8.8062	1.2345±0.2175
VorCubeRt	0.1562±0.0081	135.5521± 188.9479	1.8314±0.3755	6.0451± 7.1000	1.1621±0.1837
High energy physics citations, 27,400 nodes, 352,542 edges					
VorHalf	0.1764±0.0100	1.6620± 1.2240	1.3179±0.2909	1.6452± 1.1544	1.2107±0.2323
VorRoot	0.0611±0.0043	40.1114± 21.9262	1.9918±0.4695	11.5248± 7.9582	1.3390±0.3286
VorCubeRt	0.0461±0.0032	101.9210± 58.6233	2.0330±0.4852	9.0423± 7.5795	1.2325±0.2750
Database of Interacting Proteins, 19,928 nodes, 82,406 edges					
VorHalf	0.0117±0.0007	2.2044± 1.0637	1.1887±0.2188	2.1248± 1.0093	1.1183±0.1778
VorRoot	0.0108±0.0007	57.7343± 45.7341	1.8214±0.4084	9.1154± 6.0720	1.3216±0.3030
VorCubeRt	0.0096±0.0006	134.4816± 106.4737	1.9277±0.4444	6.2541± 3.8117	1.2644±0.2703
BioGRID, 4,039 nodes, 43,854 edges					
VorHalf	0.0035±0.0002	1.5086± 0.8003	1.2581±0.2718	1.3722± 0.6858	1.1334±0.1973
VorRoot	0.0025±0.0001	10.7295± 7.9563	1.8676±0.5737	3.0394± 1.9172	1.2753±0.3354
VorCubeRt	0.0024±0.0001	18.6805± 14.7570	1.9412±0.6250	2.7906± 1.7177	1.2308±0.3137

Every graph was preprocessed 1,000 times using different random seeds (250 times for the European road networks). For these runs we report the mean value and standard deviation of the execution time in seconds. After preprocessing, we performed 100 shortest path queries for random (s, t) pairs. For these queries, we provide the mean values and standard deviations of the speed-up relative to the bidirectional version of Dijkstra's algorithm, and of the multiplicative stretch relative to a shortest path.

6.4   Results

Running times, speed-ups, and approximation qualities for the Voronoi method are listed in Table 2, for all data sets. The performances of the other methods are listed in Table 4 as originally summarized in [SS07a, GSSD08, BDS+08].

Table 3. Road networks: This table is excerpted from Sanders and Schultes [SS07a, Table 1] except for CHHNR values, which are from [GSSD08, Table 1]. Preprocessing times are converted from minutes to seconds to ease comparison with our method. Machines used (except for A*): 2.0 or 2.6 GHz processor, 8 or 16 GB RAM, C++ implementation.

PTV European road network, driving time		prep. [s]	speed-up
CHHNR$_{EDS1235}$	[GSSD08]	602	≈8,505
A*	[GH05]	780	28
HH	[SS06]	780	4,002
HH+dist	[SS06]	900	8,320
HH+dist+A*	[DSSW06]	1,320	11,496
HNR	[SS07b]	1,440	4,079
CHHNR$_{EVSQWL}$	[GSSD08]	1,914	≈10,874
TNR-eco	[BFM+07]	2,760	471,881
TNR-gen	[BFM+07]	9,840	1,129,143

Table 4.   Public transportation networks: This table is excerpted from Bauer et al. [BDS+08, p. 13]. SHARC is evaluated in [BD08, p. 10]. The speed-up is computed according to the number of settled nodes. Machines used: 2.0 or 2.6 GHz processor, 8 or 16 GB RAM, C++ implementation.

		long distance rail		RMV		VBB	
$|V|$		1,192,736		2,277,812		2,599,953	
$|E|$		1,789,088		3,416,552		3,899,807	
		prep. [s]	speed-up	prep. [s]	speed-up	prep. [s]	speed-up
CALT-a64	[BDS+08]	87	291.84	191	267.11	123	459.30
CALT-m16	[BDS+08]	158	182.71	377	159.62	174	281.23
ALT-m16	[DW07]	291	20.30	556	18.91	604	23.04
CHHNR	[GSSD08]	286	1,620.62	2,584	2,077.69	1,636	3,124.59
CHASE	[BDS+08]	536	2,660.93	2,863	4,649.26	2,008	10,398.64
SHARC	[BD08]	12,540	81.04			36,120	118.10

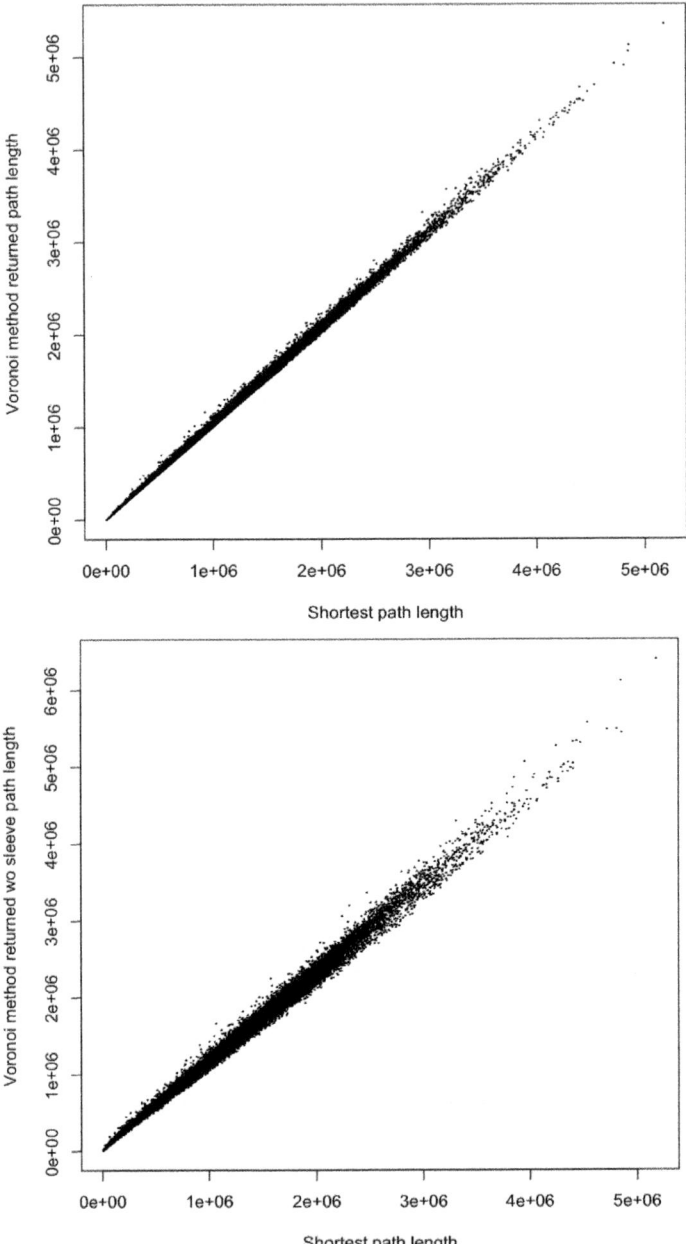

Fig. 4. Approximate path length versus actual shortest path length for VORROOT on the European road network, distance metric. Top: using sleeve. Bottom: with sleeve steps omitted. The theoretical worst-case logarithmic dependency on the number of edges cannot be observed in the experimental results. Refinement using the sleeve substantially improves the stretch in practice, although the theoretical performance is not affected.

Preprocessing. For the Voronoi method, as Lemma 3 predicts, the preprocessing cost is extremely low for all three values of p. For the non-planar graphs, the greatest preprocessing times were observed for the largest value, $p = 1/2$. This likely reflects the logarithmic cost of the heap operations associated with the computation of Voronoi regions. At the start of the Dijkstra search, the heap is initialized with all neighbors of the graph Voronoi nodes. When p is large, the initial heap size is a large proportion of the total number of nodes, and the cost of the heap operations becomes significant. On the other hand, when p and the average node degree are both small, the heap evolves smoothly with its size remaining small.

Speed-up. For road networks VORHALF achieves moderate speed-ups of approximately 2, which likely reflects the fact that the expected number of nodes of the Voronoi dual is half that of the original graph. For the power-law graphs, probability $p = 1/2$ does not lead to a significant speed-up. One reason for this might be that the Voronoi dual for each of these graphs is quite dense and, as a consequence, the Dijkstra search in the dual explores many nodes until it can find the destination. For the smaller probabilities, larger speed-ups can be observed, but the performance gain is significantly smaller than the speed-ups obtained for almost planar networks. There, the speed-up seems proportional to $1/p$. As expected, if for small values of p the sleeve is used to refine the path, the speed-up decreases drastically due to the large size of this subgraph.

Stretch. The Voronoi method achieved stretch values that were surprisingly consistent among different data sets, with most values under 2 and very close to optimal for the road networks. Figure 4 shows the approximate path length versus the shortest path length, with and without the sleeve refinement steps. The theoretical worst-case logarithmic dependency on the number of edges cannot be observed in the experimental results. Refinement using the sleeve substantially improves the stretch in practice, although the theoretical performance is not affected.

7 Conclusion

We have presented a simple and general method based on Voronoi duals to efficiently support shortest path queries in undirected graphs with very low preprocessing overheads and competitive query times, at the cost of exactness. The method was shown to be effective on a variety of graph types while remaining a reasonable alternative to existing exact methods specifically designed for transportation networks. The results of our experiments also demonstrate that the approximation ratio in practice is significantly better than the tight theoretical worst-case bound proved in the main theorem of this paper. The maximal distortion of paths in the graph Voronoi dual depends on the distance between nodes in the original graph, unlike Delaunay triangulations of the Euclidean plane, which have constant distortion [DFS90, KG92].

An interesting topic for future research would be an expected-case analysis for weighted graphs from a variety of distributions.

It remains open as to whether the Voronoi method presented in this paper can be extended to handle directed graphs. The nature of the Voronoi dual within a directed graph is inherently different from the dual within an undirected graph. The need for path connectivity suggests the construction of two Voronoi diagrams, one where reachability paths are oriented outward from Voronoi nodes and another where reachability paths are oriented inward. As the respective Voronoi regions may not coincide [Erw00], it is not straightforward to define a single dual structure whose shortest path lengths approximate those of the original graph.

References

[ACS99]   Aardal, K., Chudak, F.A., Shmoys, D.B.: A 3-approximation algorithm for the k-level uncapacitated facility location problem. Information Processing Letters 72, 161–167 (1999)

[AGK+04]   Arya, V., Garg, N., Khandekar, R., Meyerson, A., Munagala, K., Pandit, V.: Local search heuristics for k-median and facility location problems. SIAM Journal on Computing 33(3), 544–562 (2004); Announced at STOC 2001

[Aur91]   Aurenhammer, F.: Voronoi diagrams - a survey of a fundamental geometric data structure. ACM Computing Surveys 23(3), 345–405 (1991)

[BD08]   Bauer, R., Delling, D.: SHARC: Fast and robust unidirectional routing. In: Proceedings of the 10th Workshop on Algorithm Engineering and Experiments (ALENEX 2008), pp. 13–26 (2008)

[BDS+08]   Bauer, R., Delling, D., Sanders, P., Schieferdecker, D., Schultes, D., Wagner, D.: Combining hierarchical and goal-directed speed-up techniques for Dijkstra's algorithm. In: McGeoch, C.C. (ed.) WEA 2008. LNCS, vol. 5038, pp. 303–318. Springer, Heidelberg (2008)

[BFM+07]   Bast, H., Funke, S., Matijevic, D., Sanders, P., Schultes, D.: In transit to constant time shortest-path queries in road networks. In: Proceedings of the Workshop on Algorithm Engineering and Experiments (ALENEX 2007), New Orleans, Louisiana, USA, January 6 (2007)

[BGSU08]   Baswana, S., Gaur, A., Sen, S., Upadhyay, J.: Distance oracles for unweighted graphs: Breaking the quadratic barrier with constant additive error. In: Aceto, L., Damgård, I., Goldberg, L.A., Halldórsson, M.M., Ingólfsdóttir, A., Walukiewicz, I. (eds.) ICALP 2008, Part I. LNCS, vol. 5125, pp. 609–621. Springer, Heidelberg (2008)

[BK06]   Baswana, S., Kavitha, T.: Faster algorithms for approximate distance oracles and all-pairs small stretch paths. In: 47th Annual IEEE Symposium on Foundations of Computer Science (FOCS 2006), Berkeley, California, USA, October 21-24, pp. 591–602 (2006)

[Cha07]   Chan, T.M.: More algorithms for all-pairs shortest paths in weighted graphs. In: Proceedings of the 39th Annual ACM Symposium on Theory of Computing (STOC 2007), pp. 590–598 (2007)

[Coo03]   Cooperative Association for Internet Data Analysis. Router-level topology measurements (2003),
            http://www.caida.org/tools/measurement/skitter/router_topology/
            file: itdk0304_rlinks_undirected.gz

[CP07]   Chan, T.M., Patrascu, M.: Voronoi diagrams in $n \cdot 2^{O(\sqrt{\lg \lg n})}$ time. In: Proceedings of the 39th Annual ACM Symposium on Theory of Computing, San Diego, California, USA, June 11-13, pp. 31–39 (2007)

[DFS90]   Dobkin, D.P., Friedman, S.J., Supowit, K.J.: Delaunay graphs are almost as good as complete graphs. Discrete & Computational Geometry 5, 399–407 (1990)

[Dij59]   Dijkstra, E.W.: A note on two problems in connexion with graphs. Numerische Mathematik 1, 269–271 (1959)

[Dir50]   Dirichlet, G.L.: Über die Reduktion der positiven quadratischen Formen mit drei unbestimmten ganzen Zahlen. Journal für die Reine und Angewandte Mathematik 40, 209–227 (1850)

[Dor67]   Doran, J.E.: An approach to automatic problem-solving. Machine Intelligence 1, 105–124 (1967)

[DSSW06]   Delling, D., Sanders, P., Schultes, D., Wagner, D.: Highway hierarchies star. In: 9th DIMACS Implementation Challenge (2006)

[DW07]   Delling, D., Wagner, D.: Landmark-based routing in dynamic graphs. In: Demetrescu, C. (ed.) WEA 2007. LNCS, vol. 4525, pp. 52–65. Springer, Heidelberg (2007)

[EG08]   Eppstein, D., Goodrich, M.T.: Studying (non-planar) road networks through an algorithmic lens. In: Proceedings of the 16th ACM SIGSPATIAL International Symposium on Advances in Geographic Information Systems, ACM-GIS 2008, Irvine, California, USA, November 5-7, p. 16 (2008)

[EMK97]   Embrechts, P., Mikosch, T., Klüppelberg, C.: Modelling extremal events: for insurance and finance. Springer, London (1997)

[EP04]   Elkin, M., Peleg, D.: $(1 + \epsilon, \beta)$-spanner constructions for general graphs. SIAM Journal on Computing 33(3), 608–631 (2004); Announced at STOC 2001

[Erw00]   Erwig, M.: The graph Voronoi diagram with applications. Networks 36(3), 156–163 (2000)

[FT87]   Fredman, M.L., Tarjan, R.E.: Fibonacci heaps and their uses in improved network optimization algorithms. Journal of the ACM 34(3), 596–615 (1987); Announced at FOCS 1984

[FW93]   Fredman, M.L., Willard, D.E.: Surpassing the information theoretic bound with fusion trees. Journal of Computer and System Sciences 47(3), 424–436 (1993)

[GH05]   Goldberg, A.V., Harrelson, C.: Computing the shortest path: A* search meets graph theory. In: Proceedings of the Sixteenth Annual ACM-SIAM Symposium on Discrete Algorithms (SODA 2005), Vancouver, British Columbia, Canada, January 23-25, pp. 156–165 (2005)

[GKP05]   Garg, N., Khandekar, R., Pandit, V.: Improved approximation for universal facility location. In: Proceedings of the Sixteenth Annual ACM-SIAM Symposium on Discrete Algorithms (SODA 2005), Vancouver, British Columbia, Canada, January 23-25, pp. 959–960 (2005)

[GSDF03]   Getoor, L., Senator, T.E., Domingos, P., Faloutsos, C. (eds.): SIGKDD Proceedings (2003)

[GSSD08]   Geisberger, R., Sanders, P., Schultes, D., Delling, D.: Contraction hierarchies: Faster and simpler hierarchical routing in road networks. In: McGeoch, C.C. (ed.) WEA 2008. LNCS, vol. 5038, pp. 319–333. Springer, Heidelberg (2008)

[GW05]   Goldberg, A.V., Werneck, R.F.F.: Computing point-to-point shortest paths from external memory. In: Proceedings of the Seventh Workshop on Algorithm Engineering and Experiments (ALENEX 2005), pp. 26–40 (2005)

[HKRS97]   Henzinger, M.R., Klein, P.N., Rao, S., Subramanian, S.: Faster shortest-path algorithms for planar graphs. Journal of Computer and System Sciences 55(1), 3–23 (1997); Announced at STOC 1994

[KG92]   Mark Keil, J., Gutwin, C.A.: Classes of graphs which approximate the complete Euclidean graph. Discrete & Computational Geometry 7, 13–28 (1992)

[Lau04]   Lauther, U.: An extremely fast, exact algorithm for finding shortest paths in static networks with geographical background. In: Geoinformation und Mobilität – von der Forschung zur praktischen Anwendung, vol. 22, pp. 219–230 (2004)

[Ley02]   Ley, M.: The DBLP computer science bibliography: Evolution, research issues, perspectives. In: Laender, A.H.F., Oliveira, A.L. (eds.) SPIRE 2002. LNCS, vol. 2476, pp. 1–10. Springer, Heidelberg (2002)

[Meh88]   Mehlhorn, K.: A faster approximation algorithm for the Steiner problem in graphs. Information Processing Letters 27(3), 125–128 (1988)

[Mit03]   Mitzenmacher, M.: A brief history of generative models for power law and lognormal distributions. Internet Mathematics 1(2) (2003)

[Ram96]   Raman, R.: Priority queues: Small, monotone and trans-dichotomous. In: Díaz, J. (ed.) ESA 1996. LNCS, vol. 1136, pp. 121–137. Springer, Heidelberg (1996)

[Ram97]   Raman, R.: Recent results on the single-source shortest paths problem. SIGACT News 28, 81–87 (1997)

[SBR+06]   Stark, C., Breitkreutz, B.-J., Reguly, T., Boucher, L., Breitkreutz, A., Tyers, M.: Biogrid: a general repository for interaction datasets. Nucleic Acids Research 34(1), 535–539 (2006)

[Sha75]   Shamos, M.I.: Geometric complexity. In: Conference Record of Seventh Annual ACM Symposium on Theory of Computation (STOC 1975), Albuquerque, New Mexico, USA, May 5-7, pp. 224–233 (1975)

[SMS+02]   Salwinski, L., Miller, C.S., Smith, A.J., Pettit, F.K., Bowie, J.U., Eisenberg, D.: DIP, the database of interacting proteins: a research tool for studying cellular networks of protein interactions. Nucleic Acids Research 30(1), 303–305 (2002)

[SR90]   Szpankowski, W., Rego, V.: Yet another application of a binomial recurrence. Order statistics. Computing 43(4), 401–410 (1990)

[SS06]   Sanders, P., Schultes, D.: Engineering highway hierarchies. In: Azar, Y., Erlebach, T. (eds.) ESA 2006. LNCS, vol. 4168, pp. 804–816. Springer, Heidelberg (2006)

[SS07a]   Sanders, P., Schultes, D.: Engineering fast route planning algorithms. In: Demetrescu, C. (ed.) WEA 2007. LNCS, vol. 4525, pp. 23–36. Springer, Heidelberg (2007)

[SS07b]   Schultes, D., Sanders, P.: Dynamic highway-node routing. In: Demetrescu, C. (ed.) WEA 2007. LNCS, vol. 4525, pp. 66–79. Springer, Heidelberg (2007)

[Svi08]   Svitkina, Z.: Lower-bounded facility location. In: Proceedings of the Nineteenth Annual ACM-SIAM Symposium on Discrete Algorithms (SODA 2008), San Francisco, California, USA, January 20-22, pp. 1154–1163 (2008)

[Tho99]   Thorup, M.: Undirected single-source shortest paths with positive inte-
             ger weights in linear time. Journal of the ACM 46(3), 362–394 (1999);
             Announced at FOCS 1997
[Tho00a]   Thorup, M.: Floats, integers, and single source shortest paths. Journal of
             Algorithms 35(2), 189–201 (2000); Announced at STACS 1998
[Tho00b]   Thorup, M.: On RAM priority queues. SIAM Journal of Computing 30(1),
             86–109 (2000); Announced at SODA 1996
[Tho04a]   Thorup, M.: Compact oracles for reachability and approximate distances in
             planar digraphs. Journal of the ACM 51(6), 993–1024 (2004); Announced
             at FOCS 2001
[Tho04b]   Thorup, M.: Integer priority queues with decrease key in constant time
             and the single source shortest paths problem. Journal of Computer and
             System Sciences 69(3), 330–353 (2004); Announced at STOC 2003
[Tho07]   Thorup, M.: Equivalence between priority queues and sorting. Journal of
             the ACM 54(6) (2007); Announced at FOCS 2002
[TZ05]   Thorup, M., Zwick, U.: Approximate distance oracles. Journal of the
             ACM 52(1), 1–24 (2005); Announced at STOC 2001
[Vor07]   Voronoi, G.: Nouvelles applications des paramètres continus à la théorie
             des formes quadratiques. Journal für die Reine und Angewandte Mathe-
             matik 133, 97–178 (1907)
[Wil64]   Williams, J.W.J.: Algorithm 232: Heapsort. Communications of the
             ACM 7, 347–348 (1964)

On the Triangle-Perimeter Two-Site Voronoi Diagram

Iddo Hanniel[1,3] and Gill Barequet[2,3]

[1] Research Group, SolidWorks Corp.
Concord, MA 01742
ihanniel@solidworks.com
[2] Dept. of Computer Science
Tufts University
Medford, MA 02155
barequet@cs.tufts.edu
[3] Center for Graphics and Geometric Computing
Dept. of Computer Science
The Technion—Israel Institute of Technology
{iddoh,barequet}@cs.technion.ac.il

Abstract. The triangle-perimeter 2-site distance function defines the "distance" from a point x to two other points p, q as the perimeter of the triangle whose vertices are x, p, q. Accordingly, given a set S of n points in the plane, the Voronoi diagram of S with respect to the triangle-perimeter distance, is the subdivision of the plane into regions, where the region of the pair $p, q \in S$ is the locus of all points closer to p, q (according to the triangle-perimeter distance) than to any other pair of sites in S. In this paper we prove a theorem about the perimeters of triangles, two of whose vertices are on a given circle. We use this theorem to show that the combinatorial complexity of the triangle-perimeter 2-site Voronoi diagram is $O(n^{2+\varepsilon})$ (for any $\varepsilon > 0$). Consequently, we show that one can compute the diagram in $O(n^{2+\varepsilon})$ time and space.

Keywords: Distance function, planar map.

1 Introduction

The standard Voronoi Diagram of a given set of points (called sites) is a subdivision of the plane into regions, one associated with each site. Each site's region consists of all points in the plane closer to it than to any of the other sites. The Voronoi diagram has been rediscovered many times in dozens of fields of study including crystallography, geography, metrology, and biology, as well as mathematics and computer science. A comprehensive review of the various variations of Voronoi diagrams and of the hundreds of applications of them is given by Okabe, Boots, and Sugihara [7].

The notion of a 2-site Voronoi diagram of a point set S was first introduced in [2]. In this diagram, each *pair* of points $p, q \in S$ has a (possibly empty) region, that is the locus of all points in the plane closer to p, q than to any

M.L. Gavrilova et al. (Eds.): Trans. on Comput. Sci. IX, LNCS 6290, pp. 54–75, 2010.

other pair of points in S. Here, the distance is measured from a point x to a *pair* of points p, q by some 2-point distance function $\mathcal{D}(x, (p, q))$. The reference cited above investigated the combinatorial complexity of Voronoi diagrams, as well as efficient algorithms to compute them, with respect to several 2-point distance functions, such as $E(x, p) + E(x, q)$ (the *sum* of distances from x to p and q), $|E(x, p) - E(x, q)|$ (the absolute value of the difference between the two distances), etc. One of the observations in [2] was that the Voronoi diagram under the sum-of-distances 2-site function is identical to the well-known 2nd-order Voronoi diagram [3, §17] and, thus, has $O(n)$ complexity. General kth-order Voronoi diagrams of point sets, in two or higher dimensions, were also considered in the literature; see, e.g., [1,4].

We consider the triangle-perimeter 2-point distance function, $\mathcal{P}(x, (p, q))$, that measures the perimeter of the triangle whose vertices are x, p, q. (Since the roles of all of x, p, q are symmetric, we hereafter use the notation $\mathcal{P}(xpq)$.) We denote by V_S the sum-of-distances 2-point Voronoi diagram,[1] while the triangle-perimeter 2-point Voronoi diagram is denoted by $V_\mathcal{P}$.

The triangle-perimeter function is interesting since it models the amount of work spent by a client served by two suppliers, while the two suppliers need to communicate in order to serve correctly the client. For example, assume that for security, or efficiency, data are split and transmitted from a mobile phone (a point) by using two cellular antennas (sites), so that the protocol also requires the synchronization between the two antennas to ensure the correctness of the combined pieces of data. Thus, a client will prefer to be served by the two suppliers closest to it with respect to the perimeter function. This setting has many industrial and military applications. A similar construction (a graph-theoretic Voronoi diagram) was explored in the context of geographic networks [6].

The Voronoi diagram is a minimization diagram of a set of surfaces in one higher dimension. The respective Voronoi surfaces of the triangle-perimeter diagram are identical to those of the sum-of-distances diagram up to a vertical translation: the surface of the pair (p, q) in the latter diagram is raised by $|pq|$ to form the respective surface in the former diagram. Experimental results suggest that the worst-case complexity of $V_\mathcal{P}$ be similar to that of V_S. However, until now no better bound than the straightforward $O(n^{4+\varepsilon})$ upper bound[2] has been proven. On the other hand, it has been shown [2] that several 2-site Voronoi diagrams, e.g., the triangle-area diagram, have $\Theta(n^4)$ complexity.

In this paper we prove a theorem about the perimeters of triangles, two of whose vertices are on a given circle. Using this theorem and the mentioned condition, we are able to provide the upper bound $O(n^{2+\varepsilon})$ (for any $\varepsilon > 0$) on the combinatorial complexity of $V_\mathcal{P}(S)$, improving upon the previously best

---

[1] Since V_S is identical to the 2nd-order Voronoi diagram, all the properties proven in this paper for V_S apply also to the 2nd-order Voronoi diagram.

[2] The $O(n^{4+\varepsilon})$ bound comes from the combinatorial complexity of the lower envelope of $\Theta(n^2)$ "well-behaved" surfaces (in the sense of Assumptions 7.1 of [10, p. 188]). Since the perimeter functions are such surfaces and there are $O(n^2)$ site-pairs, the bound is immediate.

known bound $O(n^{4+\varepsilon})$. Recently, Dickerson and Eppstein [5] obtained the same result by a very elegant and more general proof. We also show that $V_{\mathcal{P}}(S)$ can be computed in $O(n^{2+\varepsilon})$ time and space.

Throughout the paper we use the following notation. The line-segment defined by a pair of points $p, q \in \mathbb{R}^2$ is denoted by pq, and its length is denoted by $|pq|$. The circle whose diameter is pq is denoted by $C(pq)$. Given three points $p, q, r \in \mathbb{R}^2$, we denote the triangle they induce by $\triangle pqr$. The circumscribing circle of $\triangle pqr$ is denoted by $C(pqr)$. The perimeter of the $\triangle pqr$ is denoted by $\mathcal{P}(pqr)$. On a few occasions we also write $\mathcal{P}(pqrs)$ to denote the perimeter of the quadrilateral whose vertices are the points p, q, r, s.

2   Outline of the Proof

In this section we outline the sequence of claims that form the main proof in the paper.

Our goal is to prove that if a pair of sites $p, q \in S$ does *not* have a (non-empty) face in $V_S(S)$, then neither can it have a face in $V_{\mathcal{P}}(S)$. Thus, since the complexity of $V_S(S)$ is linear in the number of sites [2, §2], the number of pairs of sites having nonempty faces in $V_{\mathcal{P}}(S)$ is also linear. This implies the bound $O(n^{2+\varepsilon})$ (for any $\varepsilon > 0$) on the complexity of $V_{\mathcal{P}}$ (Theorem 9 in Section 7).

We first recall, in Section 3, a Delaunay-like necessary and sufficient condition for a pair of sites (p, q) *not* to have a face in a V_S. This condition reduces the problem to proving that for any four points p, r, q, s ordered along a circle (see Figure 1), and for any point x in $\mathbb{R}^2$, the perimeter of at least one of $\triangle prx$, $\triangle qrx$, $\triangle psx$, and $\triangle qsx$ is smaller than the perimeter of $\triangle pqx$.

For example, it is easy to see in Figure 1 that the perimeter of $\triangle prx$ is smaller than that of the $\triangle pqx$. This is stated in the following theorem, which is the main theorem of this paper.

Theorem 1. *Given four ordered cocircular points p, r, q, s, for any point x in the plane, at least one of the following holds:*

$$(i) \quad \mathcal{P}(pqx) \geq \mathcal{P}(prx),$$
$$(ii) \quad \mathcal{P}(pqx) \geq \mathcal{P}(qrx),$$
$$(iii) \quad \mathcal{P}(pqx) \geq \mathcal{P}(psx),$$
$$(iv) \quad \mathcal{P}(pqx) \geq \mathcal{P}(qsx).$$

In order to prove Theorem 1, we break the plane into several portions and show that the claim holds for each portion separately, when the portion contains the point x. Specifically, we take the following steps:

1. In Section 4 we trim the feasible region in the plane in which the point x from Theorem 1 can exist.
   - In Theorem 3 we reduce the region to a "feasible slice" in the plane.
   - In Lemma 2 we reduce the feasible region further, to the portion outside the circle.

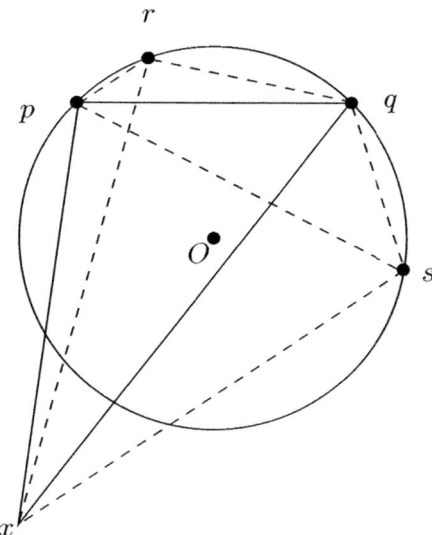

Fig. 1. Triangles induced by four points on a circle. Theorem 1 states that, for any point $x \in \mathbb{R}^2$, the minimal-perimeter triangle cannot be $\triangle pqx$.

2. In Section 5 we represent the problem using three angles, α, β, γ. We then continue the proof by shrinking the possible ranges of these angles.
   - In Theorem 5 we reduce the domain of α and prove that it cannot be greater than $\pi/6$.
   - In Theorem 6 we further reduce the domain of γ and prove that it must be between -3α and α (or, symmetrically, between $-\pi-\alpha$ and $-\pi+3\alpha$).
3. Finally, in Section 6 we conclude the proof by showing that Theorem 1 also holds for the configurations defined by the reduced domains.

After Theorem 1 is proven for a point s on $C(pqr)$, we apply in Section 6.3 an extension argument to prove the claim for points *inside* the circle. Corollary 3 concludes that if the pair (p, q) does not have a face in $V_{\mathcal{S}}(S)$, neither can it have a face in $V_{\mathcal{P}}(S)$. The complexity of $V_{\mathcal{P}}(S)$ is then proven in Theorem 9.

3   Face Condition in $V_{\mathcal{S}}(S)$

It is a well-known fact [1,8] that the faces of the 2nd-order Voronoi diagram of a point set S are in bijection with the edges in the Delaunay triangulation of S. As noted in the introduction, the former diagram is identical to $V_{\mathcal{S}}(S)$. Thus, we have:

Theorem 2. *Let S be a set of points in $\mathbb{R}^2$. The pair (p, q) has a non-empty face in $V_{\mathcal{S}}(S)$ if and only if there exists a site $r \in S \setminus \{p, q\}$ such that $C(pqr)$ is empty.*

This condition reduces our problem to that of proving the properties of quadruples of points p, q, r, s, in which s lies inside $C(pqr)$. Namely, in order to prove that if a pair of sites $(p, q) \in S$ does not have a face in $V_S(S)$, then neither can it have a face in the $V_P(S)$, we have to prove that for any point x in the plane one of the following holds:

$$
\begin{aligned}
(i) & \quad \mathcal{P}(pqx) \geq \mathcal{P}(prx), \\
(ii) & \quad \mathcal{P}(pqx) \geq \mathcal{P}(qrx), \\
(iii) & \quad \mathcal{P}(pqx) \geq \mathcal{P}(psx), \\
(iv) & \quad \mathcal{P}(pqx) \geq \mathcal{P}(qsx).
\end{aligned}
$$

Given a proof of this property for a point s on $C(pqr)$, a simple extension argument shows that the property also holds for any point s *inside* the circle (see Section 6.3). Hence, in Theorem 1 we only prove the claim for configurations of four points on the circle.

4   Reducing the Feasible Region

In this section we define a *feasible region*, such that Theorem 1 holds for any point x outside this region. Thus, we only need to prove the theorem for points inside the feasible region. In Section 4.1 we define a "feasible slice" and prove that it is a feasible region. In Section 4.2 we introduce Lemma 1, which will be used throughout the proof of the theorem, and in Section 4.3 we make use of this lemma to reduce the feasible region further to be outside the circle.

4.1   The Feasible Slice

Definition 1. *Consider the circle $C(pqr)$ defined by three sites (see Figure 2). The lines aa' and bb' are the bisectors of pr and qr, respectively, and their intersection, O, is the center of $C(pqr)$. Hence, every point to the right of aa' is closer to r than to p, and, similarly, every point to the left of bb' is closer to r than to q. Therefore, the only points in $\mathbb{R}^2$ that can be closer to (p, q) (in the sum-of-distances sense) than to (p, r) or to (q, r) are those within the (infinite) slice defined by bOa. We call this the feasible slice of p, q, r.*

The feasible slice has the following properties:

1. Its angle span is $\pi - \alpha$ (where $\alpha = \angle prq$). This can be seen by looking at the quadrilateral $Oa'rb'$, which has two right angles.
2. For any point r above (resp., below) pq, the feasible slice is bounded from the left (resp., right) by the line passing through q and O and from the right (resp., left) by the line passing through p and O. The extreme cases are attained when r coincides with p (in which case aa' passes through p and O) or when r coincides with q (in which case bb' passes through q and O).

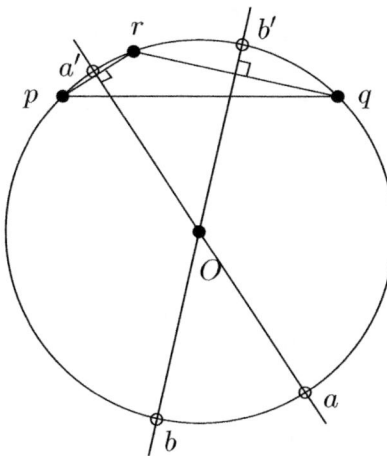

Fig. 2. The circle induced by the points p, q, r, with the bisectors of p and r, and of q and r. The infinite region bounded by the rays Ob and Oa is the feasible slice.

The properties above are true both when the point r lies on the smaller arc of the circle (as in the figure) or on the larger arc.

Observe that if (p, q) is a pair of sites that have no face in $V_S(S)$, then there must be at least one site r inside $C(pq)$. Otherwise, we could extend the circle (moving its center along the bisector of p and q) until encountering the first site r, keeping $C(pqr)$ empty, which is a contradiction to Theorem 2. This observation means that there is at least one site r such that the $\triangle pqr$ does not contain its circumcenter (the center of $C(pqr)$). We will use this fact in the proofs therein.

Theorem 3. *Let (p, q) be a pair of sites that have no face in $V_S(S)$, and let $r \in S$ be a site inside $C(pq)$.[3] For any point x outside the feasible slice of r, either (p, r) or (q, r) is closer to x than (p, q) in the triangle-perimeter sense.*

Proof. The segments pr and qr are shorter than pq since they support a shorter arc. Thus, if (without loss of generality) $|xp| + |xq| > |xp| + |xr|$, then $|xp| + |xq| + |pq| > |xp| + |xr| + |pr|$ (and similarly for q). Since, by definition, outside the feasible slice of r we have either $|xp| + |xq| > |xp| + |xr|$ or $|xp| + |xq| > |xq| + |xr|$, the claim follows.   □

4.2   Maximum Perimeter of a Chord-Based Triangle

In this section we refer, without loss of generality, to triangles whose base is parallel to the x axis and whose third vertex is above the base.

Lemma 1. *The maximum perimeter of a triangle based on a chord of a circle, whose apex lies on the circle, is achieved when it is an isosceles triangle. (The minimum is achieved by a degenerate triangle: the chord.)*

---

[3] As noted above, there is at least one such site.

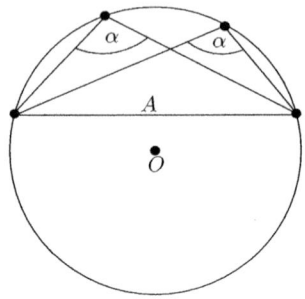

Fig. 3. Two triangles based on the same chord have the same head angle

Proof. The head angle of all triangles, whose base is the same chord, is the same (see Figure 3), so by the sine rule we have:

$$\frac{A}{\sin(\alpha)} = \frac{B}{\sin(\beta)} = \frac{C}{\sin(\pi - \alpha - \beta)}.$$

Therefore,

$$B = \frac{A\sin(\beta)}{\sin(\alpha)}, C = \frac{A\sin(\pi - (\alpha + \beta))}{\sin(\alpha)}.$$

To obtain the maximum value of $A + B + C$ we compute its derivative with respect to β (note that α and A are known) and equate the resulting term to zero:

$$(A + B + C)' = \frac{A}{\sin(\alpha)}(\sin(\pi - \alpha - \beta)) + \sin(\beta))'$$
$$= \frac{A}{\sin(\alpha)}(-\cos(\pi - \alpha - \beta)) + \cos(\beta)) = 0.$$

An elementary calculation shows that the solution (when α, β, γ are smaller than π) is $\pi - (\alpha + \beta) = \beta$, i.e., $\gamma = \beta$. Thus, the maximum is achieved when the triangle is isosceles.

The minimum value is obtained trivially from the triangle inequality. $\square$

Corollary 1. *Given two triangles whose bases are the same chord c of a circle and whose apexes are on the circle, the triangle with the larger perimeter is the one whose apex is closer to the apex of the isosceles triangle based on c.*

Proof. The claim follows from the monotonicity and symmetry of the perimeter function. $\square$

4.3   Reducing the Feasible Region to Outside the Circle

In this section we reduce the feasible region further and prove that any point x inside the circle satisfies the conditions of Theorem 1. The proof consists of two steps: In Lemma 2 we prove the claim for points on the circle itself, and then in Theorem 4 we use an extension argument to prove it for points inside the circle.

Lemma 2. *Every point x on the arc of $C(prq)$ in the feasible slice is closer to (p, r) or to (q, r) than to (p, q) in the triangle-perimeter sense.*

Proof. Consider a point x on the arc of the feasible slice, and compare the perimeters of $\triangle xqr$ and $\triangle xqp$ (see Figure 4). The apex of the isosceles triangle based on xq is the intersection of ℓ_{xq}, the bisector of xq, with the circle; denote the intersection point by ℓc_{xq}. By Corollary 1, if the point p is closer to ℓc_{xq} than r is, then we are done since in this case the perimeter of $\triangle xpq$ is greater than the perimeter of $\triangle xrq$. Therefore, we only need to analyze the case in which r is closer to ℓc_{xq} than p. We will show that in this case $\mathcal{P}(xpq) \geq \mathcal{P}(xrp)$ (i.e., the perimeter of the other triangle based on xr ($\triangle xrp$) is smaller than that of $\triangle xpq$).

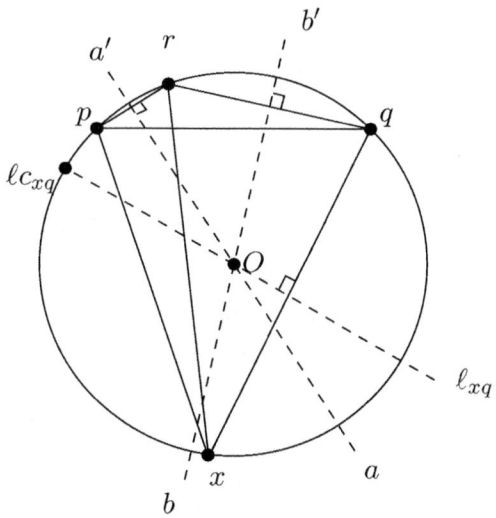

Fig. 4. A point x on the arc of the feasible slice, and the bisector line ℓ_{xq}

The fact that r is closer to ℓc_{xq} than to p means that ℓc_{xq} is to the right of a' (see Figure 4). However, in this case we can take ℓ_{xp}, the bisector of xp, in which case ℓc_{xp} must be closer to q than to r. This is because the angle between ℓ_{xp} and ℓ_{xq} is larger than $\pi/2$ (since $\angle pxq \leq \pi/2$), while the angle between aa' and bb' is smaller than $\pi/2$ (since $\angle prq \geq \pi/2$).[4]

Thus, if r is closer to ℓc_{xq} than p, it must be further from ℓc_{xp} than q, and, therefore, by Corollary 1, we have $\mathcal{P}(xpq) \geq \mathcal{P}(xrp)$. Hence, x is closer either to (p, r) or to (q, r) than to (p, q) in the triangle-perimeter sense.   □

Theorem 4. *For every point x inside or on $C(pqr)$, either $\mathcal{P}(xpr)$ or $\mathcal{P}(xqr)$ is smaller than $\mathcal{P}(xpq)$. That is, every point x inside or on $C(pqr)$ is closer to (p, r) or to (q, r) than to (p, q) in the triangle-perimeter sense.*

---

[4] The inequalities $\angle prq \geq \pi/2$ and $\angle pxq \leq \pi/2$ follow from the fact that r lies on the smaller arc on pq, on the other side of its circumcenter.

Proof. Since we know, according to Lemma 2, that the claim is true for points on the circle, we only need to prove it for points strictly inside the circle (and inside the feasible slice). We prove this by a simple argument that extends the line from r through x to the arc of the feasible slice and uses Lemma 2. Let x' be the point on $C(pqr)$ that is on the line extending from r through x. We assume, without loss of generality, that (by Lemma 2) $\mathcal{P}(prx') \leq \mathcal{P}(pqx')$.[5] Then, let $d_1 = |rx'| - |rx| = |xx'|$, $d_2 = |px'| - |px|$, and $d_3 = |qx'| - |qx|$. The difference between the perimeters $\mathcal{P}(pqx')$ and $\mathcal{P}(pqx)$ (resp., $\mathcal{P}(prx')$ and $\mathcal{P}(prx)$) is $d_2 + d_3$ (resp., $d_2 + d_1$). Since, by the triangle inequality, $d_3 \leq d_1$, and by the assumption $\mathcal{P}(prx') \leq \mathcal{P}(pqx')$, we have:

$$\mathcal{P}(pqx) = \mathcal{P}(pqx') - d_2 - d_3, \ \mathcal{P}(prx) = \mathcal{P}(prx') - d_2 - d_1.$$

Therefore,

$$\mathcal{P}(pqx) - \mathcal{P}(prx) = (\mathcal{P}(pqx') - \mathcal{P}(prx')) + (d_1 - d_3) \geq 0,$$

i.e., $\mathcal{P}(prx) \leq \mathcal{P}(pqx)$. □

5   Angle Representation and Reducing the Range of Angles

As noted in Section 3, our proof is reduced to proving properties of a configuration of four points p, q, r, s on a circle. We assume, without loss of generality, that this is a unit circle.

The points p, q, r, s can be defined by using three angles:
$$p = (-\cos(\alpha), \sin(\alpha)),$$
$$q = (\cos(\alpha), \sin(\alpha)),$$
$$r = (\cos(\beta), \sin(\beta)),$$
$$s = (\cos(\gamma), \sin(\gamma)),$$
where $\alpha < \beta < \pi - \alpha$ and $-\pi - \alpha < \gamma < \alpha$.

In the next theorem we prove that $\alpha \leq \pi/6$ by showing that if $\alpha > \pi/6$, then one of the conditions (i–iv) of Theorem 1 is met. In Section 5.1 we further reduce the angle γ and prove that it must be between -3α and α (or, symmetrically, between $-\pi - \alpha$ and $-\pi + 3\alpha$).

Theorem 5. $\alpha \leq \pi/6$.

We now sketch a proof of Theorem 5 which is based on transforming it to a set of equations in α and β and tracing the space of solutions. In the appendix we provide a more complete analytical (albeit more complex) proof of the theorem. The main idea of the proof is showing that if $\alpha > \pi/6$, then, for any point x in the feasible region, the perimeter of either $\triangle xpr$ or $\triangle xqr$ is smaller than that of $\triangle xpq$. Thus, in order to satisfy Theorem 1, α cannot exceed $\pi/6$. We do so by reformulating the problem in terms of bounding hyperbolae.

---

[5] Otherwise, $\mathcal{P}(qrx') \leq \mathcal{P}(pqx')$ and an identical argument holds.

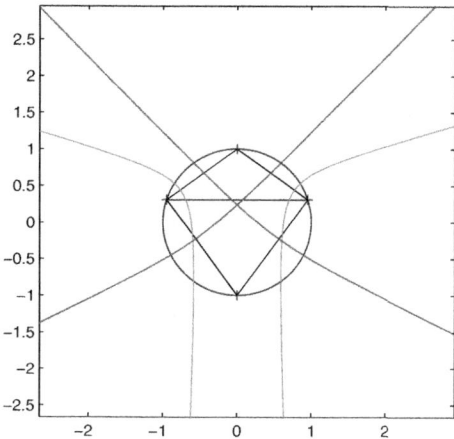

Fig. 5. The hyperbolae induced by the perimeters of p, q, r, s. H_{pr} and H_{qr} (on the upper segments) and H_{ps} and H_{qs} (on the lower segments) for $\alpha = 0.1\pi$, $\beta = 0.5\pi$, and $\gamma = -0.5\pi$.

An alternative way to view the problem described in Theorem 1 is as the intersection of four portions of the planes bounded by hyperbolae. Proving Theorem 1 then reduces to proving that the conjunction of the following conditions is empty (see Figure 5 for an illustration):

H_{pr} : $|px| - |rx| \leq |qr| - |pq|$ (induced by $|qx| + |px| + |pq| \leq |qx| + |rx| + |qr|$);
H_{qr} : $|qx| - |rx| \leq |pr| - |pq|$ (induced by $|qx| + |px| + |pq| \leq |px| + |rx| + |pr|$);
H_{ps} : $|px| - |sx| \leq |qs| - |pq|$ (induced by $|qx| + |px| + |pq| \leq |qx| + |sx| + |qs|$);
H_{qs} : $|qx| - |sx| \leq |ps| - |pq|$ (induced by $|qx| + |px| + |pq| \leq |px| + |sx| + |ps|$).

In terms of the hyperbola formulation, Theorem 5 means that if $\alpha > \pi/6$, then the hyperbolae H_{pr} and H_{qr} do not intersect. In order to prove this, we need to show that the head angle of $\triangle pqr$ (which is $\pi + \alpha$) is larger than the sum of the two asymptote angles ($\theta_1 + \theta_2$).

The asymptote angle (between the base segment and the asymptotes of the hyperbola) can be written as $\theta = \operatorname{atan}(\sqrt{\frac{c^2 - a^2}{a^2}})$, where c is the length of the base segment (e.g., $|pr|$ for H_{pr}) and a is the quantity on the right side of the hyperbola inequality (e.g., $|qr| - |pq|$ for H_{pr}).

Since $|pr|$, $|qr|$, and $|pq|$ can be formulated as functions of α and β, it is sufficient to show that the function $(\theta_1(\alpha, \beta) + \theta_2(\alpha, \beta))$ is smaller than $\pi + \alpha$ for any $\alpha > \pi/6$.

Figure 6 shows the sum-of-asymptote-angles function $F(\alpha, \beta) = (\theta_1(\alpha, \beta) + \theta_2(\alpha, \beta))$ and its intersection with the plane $\pi + \alpha$, the head-angle function $(0 \leq \alpha \leq \pi/2, \alpha \leq \beta \leq \pi/2 - \alpha)$. For $\alpha > \pi/6$ (in fact, even for $\alpha > 0.1\pi$), it is evident that the sum of asymptote angles is smaller than the head angle, and, therefore the hyperbolae do not intersect.

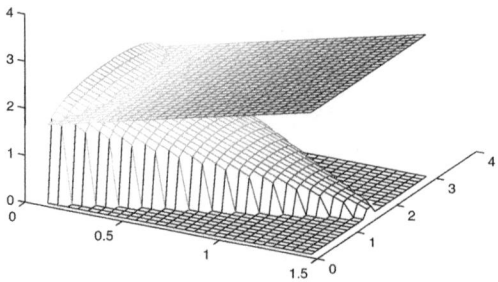

Fig. 6. The intersection of the head-angle plane $\pi+\alpha$ and the sum-of-asymptote-angles function. For $\alpha > 0.5$, the head angle is greater than the sum of asymptote angles.

5.1   Constraining γ

Theorem 6. $-3\alpha \leq \gamma \leq \alpha$ (or, symmetrically, $(-\pi - \alpha) \leq \gamma \leq -\pi + 3\alpha$).

Proof. If $\alpha \leq \pi/6$ and $-\pi + 3\alpha \leq \gamma \leq -3\alpha$, we have both $|ps| \leq |pq|$ and $|qs| \leq |pq|$ (see Figure 7). Since, by our assumption, (p, q) does not have a face in V_S, we either have $|qx| + |px| \geq |qx| + |sx|$ or $|qx| + |px| \geq |px| + |sx|$. However, if $|qx| + |px| \geq |qx| + |sx|$ (resp., $|qx| + |px| \geq |px| + |sx|$), then $|qx| + |px| + |pq| \geq |qx| + |sx| + |qs|$ (resp., $|qx| + |px| + |pq| \geq |px| + |sx| + |ps|$). Therefore, for $(-\pi + 3\alpha) \leq \gamma \leq -3\alpha$, we either have $\mathcal{P}(pqx) \geq \mathcal{P}(psx)$ or $\mathcal{P}(pqx) \geq \mathcal{P}(qsx)$, so that either condition (iii) or (iv) of Theorem 1 holds.   □

Corollary 2. *If s is on the right half of the plane (i.e., $-3\alpha \leq \gamma \leq \alpha$), then x cannot be on the same side of the ps-bisector as s (the lower right side—see Figure 8). Symmetrically, if s is on the left half of the plane ($-\pi - \alpha \leq \gamma \leq -\pi + 3\alpha$), then x cannot be on the same side as s (the lower left side) of the qs-bisector.*

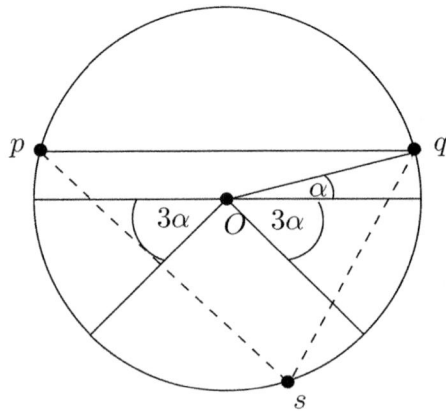

Fig. 7. If $\alpha \leq \pi/6$ and $\pi + 3\alpha \leq \gamma \leq -3\alpha$, we have both $|ps| \leq |pq|$ and $|qs| \leq |pq|$.

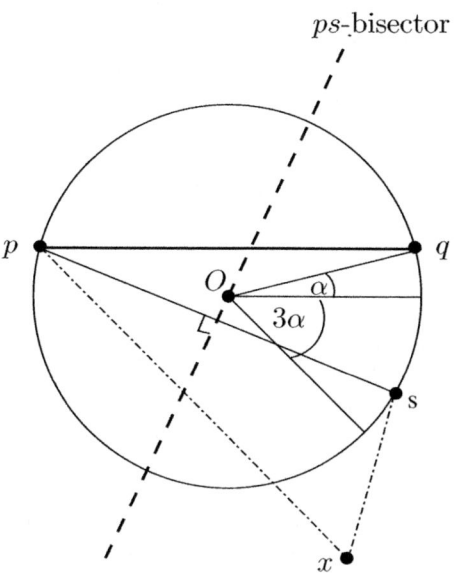

Fig. 8. If x is on the same side of the ps-bisector as s, then $\mathcal{P}(xqs) < \mathcal{P}(xpq)$

Proof. If $-3\alpha \leq \gamma \leq \alpha$, then $|sq| \leq |pq|$. Furthermore, if x is closer to s than to p (on the same side of the ps-bisector as s, see Figure 8), then we again have $|xs| < |xp|$, and, thus, $\mathcal{P}(xqs) = |sq| + |xs| + |qx| < |pq| + |xp| + |qx| = \mathcal{P}(xpq)$ and (iv) hold. A similar argument holds for the symmetric case.   □

6   Concluding the Proof

In this section we conclude the proof of the main theorem. In what follows we refer to the center of $C(pqs)$ as the center of coordinates, and assume, without loss of generality, that the segment pq is parallel to the x axis and lying above it. The "right" and "left" (resp., "upper" and "lower") halves of the plane are with respect to the y (resp., x) axis. The final proposition is as follows:

Theorem 7. *Under the above constraints on α and on the feasible region, if s is on the right half of the plane (i.e., $-3\alpha \leq \gamma \leq \alpha$), then for any point x on the left half of the plane either $\mathcal{P}(prx) \leq \mathcal{P}(pqx)$ or $\mathcal{P}(psx) \leq \mathcal{P}(pqx)$, i.e., either condition (i) or (iii) of Theorem 1 holds (and, hence, Theorem 1 is proven).*

Note that we actually need to prove the claim only for the third (lower left) quarter and outside the unit circle, because the feasible region is outside the unit circle and contained in the lower half of the plane (since it is delimited by the bisectors of pr and qr). Also note that from Corollary 2 we have that Theorem 7 is sufficient to finish the proof of the main theorem, since if s is in the right (resp., left) half of the plane, x cannot be on the right (resp., left) half of the plane. Therefore, it is sufficient to prove that x cannot be on the left (resp., right) half of the plane in order to conclude our proof.

The proof of Theorem 7 consists of two steps. In Lemma 3 we use Corollary 1 to prove that for a point x on the third-quarter arc, either $\mathcal{P}(prx) < \mathcal{P}(pqx)$ or $\mathcal{P}(psx) < \mathcal{P}(pqx)$. We then use an extension argument to generalize the result to any point in the third quarter that is outside the circle, and, thus, prove Theorem 7.

6.1   A Point on the Circle

Lemma 3. *Let $\ell_{2\alpha q}$ be the line passing through q at angle 2α with the positive x axis (see Figure 9). Under the above constraints on (p, q) ($\alpha \le \pi/6$) and on s ($-3\alpha \le \gamma \le \alpha$), for any point x on the arc in the third quarter, we have the following: If x is above the line $\ell_{2\alpha q}$, then $\mathcal{P}(prx) < \mathcal{P}(pqx)$, and if x is below the line, then $\mathcal{P}(psx) < \mathcal{P}(pqx)$.*

Proof. If x is below $\ell_{2\alpha q}$ (and on the arc in the third quarter), then ℓc_{xp} (the intersection of the circle and ℓ_{xp}, (the bisector of px, see Figure 9) is in the first quarter above q (ℓ_{xp} passes through q when x is on the line $\ell_{2\alpha q}$). Thus, by Corollary 1, since q is closer to ℓc_{xp} than s, $\mathcal{P}(psx) < \mathcal{P}(pqx)$. If x is above $\ell_{2\alpha q}$, then the point ℓc_{xp} is below q. Thus, by Corollary 1, since q is closer to ℓc_{xp} than r, we have $\mathcal{P}(prx) < \mathcal{P}(pqx)$. □

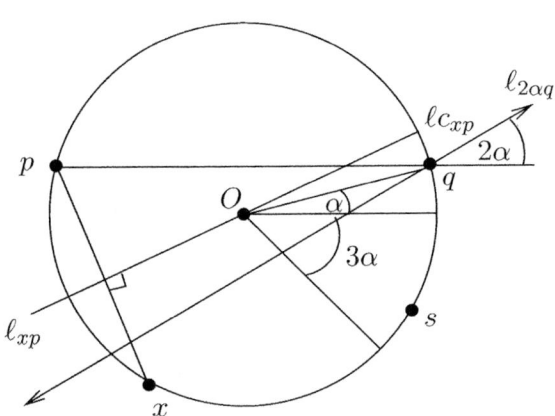

Fig. 9. The line $\ell_{2\alpha q}$ passing through q at angle 2α

6.2   Extending Theorem 7 to External Points

We are now ready to prove Theorem 7 for points outside the circle (points inside the circle are not in the feasible region, as was proven in Theorem 4). The proof is based on an extension argument.

Proof. (Theorem 7) We prove that under the above constraints on (p, q) and s, for any point x in the third quarter (and outside the circle), if x is above the line $\ell_{2\alpha q}$, then $\mathcal{P}(pxr) < \mathcal{P}(pxq)$, and if x is below the line, then $\mathcal{P}(pxs) < \mathcal{P}(pxq)$.

Extend x to q and let x' be the point of intersection with the circle (see Figure 10). We need to consider two cases:

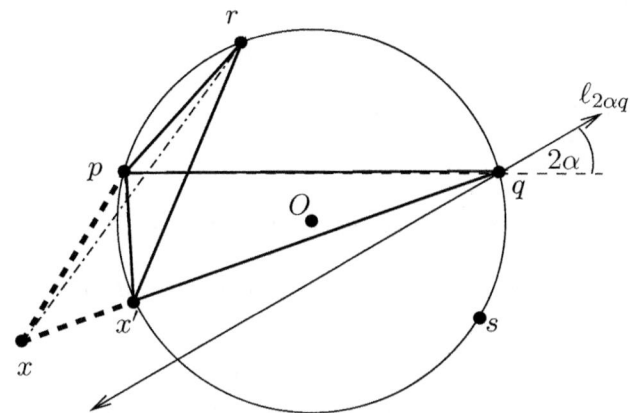

Fig. 10. Proving Theorem 7 using the point x'

(i) x is above the line. From Lemma 3 we have that $\mathcal{P}(px'r) < \mathcal{P}(px'q)$. Therefore, if we remove px' from $\triangle px'r$ and $\triangle px'q$, and replace it them by the two segments px and xx', we obtain the quadrilateral $pxx'r$ and the triangle $pxx'q = \triangle pxq$. Having only substituted $|px'|$ by $|px|+|xx'|$, it readily follows that $\mathcal{P}(pxx'r) < \mathcal{P}(pxx'q) = \mathcal{P}(pxq)$. Since $|xr| < |xx'| + |x'r|$, we obtain (by the triangle inequality) that $\mathcal{P}(pxr) < \mathcal{P}(pxx'r)$. Thus,

$$\mathcal{P}(pxr) < \mathcal{P}(pxx'r) < \mathcal{P}(pxx'q) = \mathcal{P}(pxq).$$

(ii) x is below the line. A similar argument shows that

$$\mathcal{P}(pxs) < \mathcal{P}(pxx's) < \mathcal{P}(pxx'q) = \mathcal{P}(pxq). \qquad \square$$

6.3   Extending Theorem 7 to Internal Points

Having proved Theorem 1 for four sites on the circle, we now use an extension argument to prove the claim for a point s inside $C(pqr)$.

Theorem 8. *Given three cocircular points p, r, q, where $|pq|$ is smaller than the diameter of $C(pqr)$, and a point s inside $C(pqr)$, where s and r are separated by the line supporting pq, for any point x in the plane, at least one of the following holds: $\mathcal{P}(pqx) \geq \mathcal{P}(prx)$, $\mathcal{P}(pqx) \geq \mathcal{P}(qrx)$, $\mathcal{P}(pqx) \geq \mathcal{P}(psx)$, or $\mathcal{P}(pqx) \geq \mathcal{P}(qsx)$.*

Proof. If s lies inside $\triangle pqx$, then it is trivial that $\mathcal{P}(pqx) \geq \mathcal{P}(psx)$ and $\mathcal{P}(pqx) \geq \mathcal{P}(qsx)$. Thus, we only need to consider the case in which s is inside $C(pqr)$ and outside $\triangle pqx$. In this case, we extend the line-segment xs on the side of s until it intersects $C(pqr)$ again. Let s' denote the intersection point. Then, from Theorem 1 we know that $\mathcal{P}(pqx) \geq \mathcal{P}(ps'x)$ or $\mathcal{P}(pqx) \geq \mathcal{P}(qs'x)$. Since $|xs| < |xs'|$, we have that $\mathcal{P}(ps'x) > \mathcal{P}(psx)$ and $\mathcal{P}(qs'x) > \mathcal{P}(qs'x)$, and the claim follows. □

Combining Theorem 8 with Corollary 2 results in the following:

Corollary 3. *If a pair of sites (p, q) has no face in $V_S(S)$, then neither can it have a face in $V_{\mathcal{P}}(S)$.*

7   Complexity and Computation of $V_{\mathcal{P}}(S)$

Theorem 9. *Let S be a set of n points. Then, the combinatorial complexity of $V_{\mathcal{P}}(S)$ is $O(n^{2+\varepsilon})$ (for any $\varepsilon > 0$).*

Proof. From Corollary 3 we know that a pair of points in S has a non-empty region in $V_{\mathcal{P}}(S)$ only if this pair has a non-empty region in the 2-site Voronoi diagram under the sum-of-distances function. As already noted in [2], the latter diagram is identical to the regular 2nd-order Voronoi diagram, whose complexity is known to be $\Theta(n)$. This means that only $O(n)$ pairs of sites from S have non-empty regions in $V_{\mathcal{P}}(S)$. This by itself is not enough to guarantee the linear complexity of $V_{\mathcal{P}}(S)$, since a single region may be broken into many cells in the diagram.[6] However, we can easily show a slightly-superquadratic bound on the complexity of the diagram. Each pair that has a non-empty region in the diagram induces a Voronoi surface whose function is simply the perimeter of the respective triangle. These are "well-behaved" functions (in the sense of Assumptions 7.1 of [10, p. 188]), and so, the combinatorial complexity of their lower envelope is $O(n^{2+\varepsilon})$ (for any $\varepsilon > 0$). Since $V_{\mathcal{P}}(S)$ is the orthogonal projection of this lower envelope, the claim follows. □

Theorem 10. *Let S be a set of n points in the plane. $V_{\mathcal{P}}(S)$ can be computed in $O(n^{2+\varepsilon})$ time and space, for any $\varepsilon > 0$.*

Proof. As noted in Section 3, the Voronoi surfaces relevant for computing $V_{\mathcal{P}}(S)$ can easily be identified by the edges of the Delaunay triangulation of S. $V_{\mathcal{P}}(S)$ can be computed by applying the general divide-and-conquer algorithm of [10, pp. 202–203] for computing the lower envelope of a collection of bivariate functions. The merging step of this algorithm uses the standard line-sweep procedure of [9]. The total running time of the algorithm is $O((M + M_1 + M_2) \log N)$, where M, M_1, and M_2 are the complexities of the envelope and the two subenvelopes, respectively, and N is the number of surfaces. In our case $M = M_1 = M_2 =$

---

[6] We were able to construct examples in which a single pair of sites has $\Theta(n)$ cells in $V_{\mathcal{P}}(S)$.

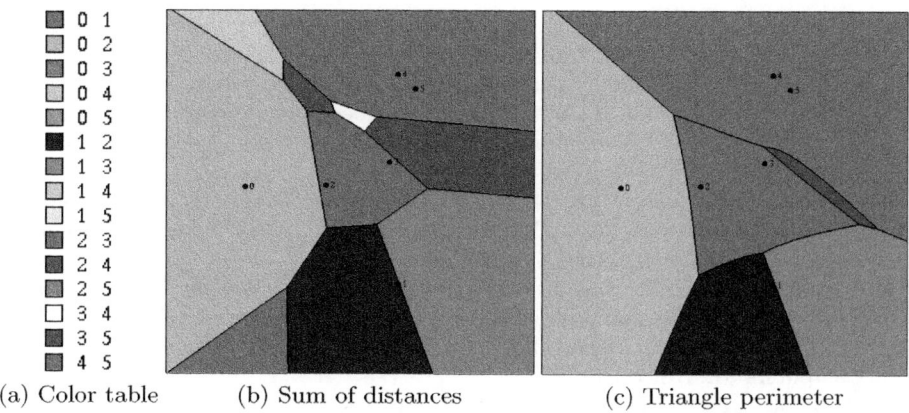

(a) Color table   (b) Sum of distances   (c) Triangle perimeter

Fig. 11. The sum-of-distances and perimeter-distance Voronoi diagrams of six points

$O(n^{2+\varepsilon})$ and $N = O(n)$, so one can compute $V_{\mathcal{P}}(S)$ in $O(n^{2+\varepsilon})$ time. The space required by the algorithm is dominated by the output size. Therefore, the algorithm requires $O(n^{2+\varepsilon})$ space.   □

8   Conclusion

We have proved a theorem on the relations between perimeters of triangles, two of whose vertices lie on a circle. Consequently, we improved the upper bound on $|V_{\mathcal{P}}(S)|$ to $O(n^{2+\varepsilon})$, for any $\varepsilon > 0$. It easily followed that one can compute $V_{\mathcal{P}}(S)$ in $O(n^{2+\varepsilon})$ time and space.

In fact, we have been able to construct an n-point set S in which the complexity of the region of a *single* pair of sites in $V_{\mathcal{P}}(S)$ is $\Theta(n)$. Furthermore, in this example, the mentioned cell consists of $\Theta(n)$ connected components (cells in the diagram). However, we were not able to construct examples with many such complex regions. Thus, we conjecture that the combinatorial complexity of $|V_{\mathcal{P}}(S)|$ is subquadratic. A sample 6-point set and its 2-site sum-of-distances and triangle-perimeter Voronoi diagrams are shown in Figure 11. (The diagram $V_S(S)$ is incomplete since the face of the pair (1,5), which lies beyond the right boundary of the figure, is missing.)

References

1. Aurenhammer, F.: A new duality result concerning Voronoi diagrams. Discrete & Computational Geometry 5, 243–254 (1990)
2. Barequet, G., Dickerson, M.T., Drysdale, R.L.S.: 2-point site Voronoi diagrams. Discrete Applied Mathematics 122, 37–54 (2002)
3. Boissonnat, J.-D., Yvinec, M.: Algorithmic Geometry (Translated from French by Hervé Brönnimann). Cambridge University Press, Cambridge (1998)

4. Chazelle, B., Edelsbrunner, H.: An improved algorithm for constructing kth-order Voronoi diagrams. IEEE Trans. on Computing C36, 1349–1354 (1987)
5. Dickerson, M.T., Eppstein, D.: Animating a continuous family of two-site Voronoi diagrams (and a proof of a bound on the number of regions). In: Video Review at the 25th Ann. ACM Symp. on Computational Geometry, Aarhus, Denmark, pp. 92–93 (2009)
6. Dickerson, M.T., Goodrich, M.T.: Two-site Voronoi diagrams in geographic networks. In: Proc. 16th ACM SIGSPATIAL Int. Conf. on Advances in Geographic Information Systems, Irvine, CA (2008)
7. Okabe, A., Boots, B., Sugihara, K.: Spatial Tessellations: Concepts and Applications of Voronoi Diagrams. John Wiley & Sons, Chichester (1992)
8. Lee, D.T.: On k-nearest neighbor Voronoi diagrams in the plane. IEEE Trans. on Computers 31, 478–487 (1982)
9. Preparata, F.P., Shamos, M.I.: Computational Geometry: An Introduction. Springer, Heidelberg (1985)
10. Sharir, M., Agarwal, P.K.: Davenport-Schinzel Sequences and Their Geometric Applications. Cambridge University Press, Cambridge (1995)

Appendix: Proof of Theorem 5

We will prove that for $\alpha > \pi/6$, for any point x in the feasible region, the perimeter of either $\triangle xpr$ or $\triangle xqr$ is smaller than that of $\triangle xpq$. That is, for any point x we have either $(|xp| + |xr| + |pr|) - (|xp| + |xq| + |pq|) < 0$ or $(|xq| + |xr| + |qr|) - (|xp| + |xq| + |pq|) < 0$. Thus, we need to prove the difference inequalities:

$$(i) \quad (|xr| + |pr|) - (|xq| + |pq|) < 0, \ or$$
$$(ii) \quad (|xr| + |qr|) - (|xp| + |pq|) < 0.$$

The sketch of the proof is as follows:

1. In Lemma 4 we first show that for any point x in the feasible region, the left-hand side of inequality (i) increases as we go to the right and the left-hand side of inequality (ii) increases as we go to the left. (Formally, for any constant size of $|xr|$, rotating xr clockwise increases the left-hand side of (ii) and decreases the left-hand side of (i).) Thus, it is sufficient to prove our proposition for x such that (i)=(ii), i.e., for any x for which $(|xr| + |pr|) - (|xq| + |pq|) = (|xr| + |qr|) - (|xp| + |pq|)$, since then for any $x' \neq x$ either (i) or (ii) will hold.

   This geometric place is the hyperbola $|xp| - |xq| = |qr| - |pr|$ and, thus, our proof is reduced to proving that, for $\alpha > \pi/6$ and for any given r, for any x on the hyperbola (ii) holds (recall that, by definition, (i)=(ii) on the hyperbola), that is,

$$(ii) \ (|xr| + |qr|) - (|xp| + |pq|) < 0,$$

given $|xp| - |xq| = |qr| - |pr|$.

2. In Lemma 5 we reformulate (ii) in terms of the angle $\angle prx$ (we do so by projecting p onto xr and achieving a bound on the difference $|xr| - |xp|$) and show that its left-hand side reaches its supremum at the hyperbola asymptotes. Thus, it is sufficient to prove our proposition for the asymptotes (i.e., when the line rx is parallel to the asymptote).

3. In Lemma 6 we analyze the asymptote angle and compare it to $\pi - \beta$, which, by definition, is the slope of the line ro. We show that for $\alpha > \pi/6$, $\pi - \beta$ is smaller than the asymptote angle θ. Consequently, $(|xr|+|qr|)-(|xp|+|pq|) = |qr| - |pq| + (|xr| - |xp|) < |qr| - |pq| + |rp_{ro}|$, where $|rp_{ro}|$ is the distance between r and p_{ro}, the projection of p onto ro. We do this since β is easier to handle than θ. Thus, all that is left to prove is that $|qr| - |pq| + |rp_{ro}| < 0$ for $\alpha > \pi/6$.

4. In Lemma 7 we analyze the function $h(\beta) = |qr| - |pq| + |rp_{ro}|$ and prove that $|qr| - |pq| + |rp_{ro}| < 0$ for $\alpha > \pi/6$, and, thus, our proof is completed.

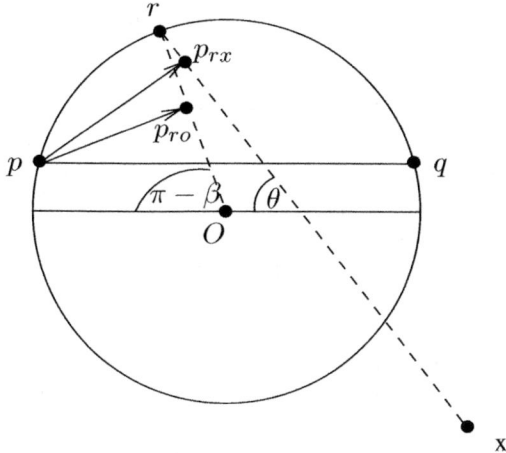

Fig. 12. The distance between r and p_{ro}, the projection of p onto ro, is greater than the distance between r and p_{rx}, the projection of p onto rx. See Corollary 6.

Figure 12 demonstrates one of the main insights of the proof. It shows the projections of p onto ro and onto rx. In Corollary 6 we prove that the distance between r and p_{ro} (the projection of p onto ro) is larger than the distance between r and p_{rx} (the projection of p onto rx).

Lemma 4. *For any constant size of xr, rotating x clockwise around r increases $(|xr| + |qr|) - (|xp| + |pq|)$ (the left-hand side of (ii)) and decreases $(|xr| + |pr|) - (|xq|+|pq|)$ (the left-hand of (i)). On the other hand, rotating xr counterclockwise decreases $(|xr| + |qr|) - (|xp| + |pq|)$ (the left-hand side of (ii)) and increases $(|xr| + |pr|) - (|xq| + |pq|)$ (the left-hand side of (i)).*

Proof. Rotating xr clockwise increases $|xq|$ and decreases $|xp|$ since the angle $\angle prx$ decreases and the angle $\angle qrx$ increases. Therefore, since $|xr|$, $|qr|$, and $|pq|$ are not changed, the result follows. □

Corollary 4. *If (i) and (ii) hold for x such that (i)=(ii), then (i) or (ii) holds for any place in the feasible region with the same distance $|xr|$.*

Proof. If xr is rotated clockwise about r, then (i) holds, and if xr is rotated counterclockwise, then (ii) holds. □

Therefore, it is sufficient to prove our proposition for x such that $(|xr| + |pr|) - (|xq| + |pq|) = (|xr| + |qr|) - (|xp| + |pq|)$, since for x to its left or right either (i) or (ii) holds. Thus, we focus on proving our proposition for any x for which $(|xr| + |pr|) - (|xq| + |pq|) = (|xr| + |qr|) - (|xp| + |pq|)$. This geometric place is the hyperbola $|xp| - |xq| = |qr| - |pr|$ and, thus, our proof is reduced to proving that, for $\alpha > \pi/6$ and for any given r, for any x on the hyperbola (ii) holds, i.e.,

$$(ii)\ (|xr| + |qr|) - (|xp| + |pq|) < 0,$$

given $|xp| - |xq| = |qr| - |pr|$.

We assume hereafter on that both $|xp|$ and $|xq|$ are smaller than $|xr|$. Otherwise, (ii) or (i) hold trivially since $|qr|$ and $|pr|$ are smaller than $|pq|$.

Let us look at the formula $(|xr| + |qr|) - (|xp| + |pq|)$. By rearranging the terms we have $(|xr| - |xp|) + (|qr| - |pq|)$. The second term is constant (for a given r) and so we are interested in bounding the first term $(|xr| - |xp|)$ from above. Let ϕ be the angle $\angle prx$ and let p_{rx} be the projection point of p onto xr. Then, the difference $|xr| - |xp|$ is bounded by the distance $|rp_{rx}|$, i.e., $|xr| - |xp| \leq |rp_{rx}|$. Equality is achieved only when x tends to infinity.

We assume without loss of generality that x is on the right half-plane, which means that r is on the left half-plane (i.e., $\beta \geq \pi/2$). The symmetric case can be proven in the same manner.

Lemma 5. *For x on the hyperbola $|xp| - |xq| = |qr| - |pr|$, the value of $(|xr| + |qr|) - (|xp| + |pq|)$ reaches its supremum at the asymptotes of the hyperbola. The supremum value of $(|xr| - |xp|)$ is then $(|xr| - |xp|) = |rp_{rx}| = |pr|\cos(\phi)$.*

Proof. Since $\triangle prp_{rx}$ is a straight angle triangle, $|rp_{rx}| = |pr|\cos(\phi)$. As x tends downwards towards infinity along the hyperbola (as x goes "down" along the hyperbola), ϕ decreases and, thus, $|rp_{rx}| = |pr|\cos(\phi)$ increases. Furthermore, only at infinity does the equality $(|xr| - |xp|) = |rp_{rx}|$ hold. Thus, the supremum of $(|xr| + |qr|) - (|xp| + |pq|)$ is reached when x tends to infinity on the hyperbola, and then $(|xr| - |xp|) = |rp_{rx}| = |pr|\cos(\phi)$. □

Corollary 5. *If $|pr|\cos(\phi) + |qr| - |pq| < 0$ then $(|xr| + |qr|) - (|xp| + |pq|) < 0$ (i.e., (ii) holds).*

Proof. By Lemma 5, $(|xr| + |qr|) - (|xp| + |pq|) \leq |pr|\cos(\phi) + |qr| - |pq|$. □

As a result from this corollary, it is sufficient to prove that for any r (for $\alpha \geq \pi/6$) we have $|pr| \cos(\phi) + |qr| - |pq| < 0$, where ϕ is the angle between the line pr and the asymptote of the hyperbola $|xp| - |xq| = |qr| - |pr|$. The angle ϕ, however, is difficult to formulate in terms of α and β. Therefore, we wish to bound it from below and so obtain an upper bound on $|pr| \cos(\phi)$. We will prove that for $\alpha \geq \pi/6$, ϕ is bounded from below by the angle $\angle pro$ (o is the origin and, thus, the slope of the line ro is $\pi - \beta$). We prove this by formulating the angle of the asymptote in terms of α and β, and showing that it is smaller than $\pi - \beta$ for $\alpha \geq \pi/6$.

Lemma 6. *For $\alpha \geq \pi/6$, $\phi \geq \angle pro$.*

Proof. Let θ be the angle between the asymptote and the x-axis (we assume without loss of generality that r is on the left half-plane, i.e., $\beta \geq \pi/2$ and so is the induced angle θ, the argument is symmetric for $\beta \leq \pi/2$). We will show that $\theta \leq \pi - \beta$ (the slope of ro) and since both ro and rx go through r, this immediately means that $\angle prx = \phi \geq \angle pro$.

The proof is based on comparing the tangent formulas of the asymptote with $\tan(\pi - \beta)$. The tangent of the asymptote of the hyperbola $|xp| - |xq| = |qr| - |pr|$ is given by

$$\sqrt{\frac{|pq|^2}{(|qr| - |pr|)^2} - 1} = \sqrt{\frac{\cos^2(\alpha)}{[\sin(\frac{\beta - \alpha}{2}) - \cos(\frac{\beta + \alpha}{2})]^2} - 1}.$$

We wish to prove that

$$\sqrt{\frac{\cos^2(\alpha)}{[\sin(\frac{\beta - \alpha}{2}) - \cos(\frac{\beta + \alpha}{2})]^2} - 1} \leq |\tan(\pi - \beta)|.$$

By squaring both sides (both sides are positive) and rearranging terms we see that

$$\frac{\cos^2(\alpha)}{[\sin(\frac{\beta - \alpha}{2}) - \cos(\frac{\beta + \alpha}{2})]^2} - 1 - \tan^2(\beta) \leq 0,$$

or, equivalently,

$$\frac{\cos^2(\alpha)}{[\sin(\frac{\beta - \alpha}{2}) - \cos(\frac{\beta + \alpha}{2})]^2} - \frac{1}{\cos^2(\beta)} \leq 0.$$

Thus, all that remains to show is that the function

$$f(\beta) = \cos^2(\alpha) \cos^2(\beta) - [\sin(\frac{\beta - \alpha}{2}) - \cos(\frac{\beta + \alpha}{2})]^2$$

is never positive (for $\alpha \geq \pi/6$ and $\pi/2 \leq \beta \leq \pi - \alpha$). Since

$$f'(\beta) = -\cos^2(\alpha) \sin(2\beta) - \sin(\alpha) \cos(\beta) - \cos(\beta)$$
$$= -\cos(\beta)[2 \sin(\beta) \cos^2(\alpha) + \sin(\alpha) + 1],$$

we have that $f'(\beta) = 0$ either when $\beta = \pi/2$ or when $\frac{1+\sin(\alpha)}{-2\cos^2(\alpha)} = \sin(\beta)$.

For $\alpha > \pi/6$ the latter equation has no solution (since $|\frac{1+\sin(\alpha)}{-2\cos^2(\alpha)}| > 1$). Thus, $f(\beta)$ is monotone and the maximum is achieved at $\beta = \pi/2$ (it is easy to verify that $\beta = \pi - \alpha$ is the minimum).

Since for $\beta = \pi/2$ we have $\theta = \pi - \beta$, we have that for $\beta > \pi/2$, $\tan(\theta) < \tan(\pi - \beta)$ and, thus, $\angle prx = \phi \geq \angle pro$. □

Corollary 6. Let p_{ro} be the projection of p onto ro. For $\alpha > \pi/6$, the distance $|rp_{rx}| \leq |rp_{ro}|$, i.e., the distance between r and the projection of p onto rx is smaller than the distance between r and the projection of p onto ro (see also Figure 12).

Proof. Since $\phi \geq \angle pro$, we have $|rp_{rx}| = |pr|\cos(\phi) \leq |pr|\cos(\angle pro) = |rp_{ro}|$. □

Therefore, by Corollary 5, if we prove that $|rp_{ro}| + |qr| - |pq| < 0$ for any r when $\alpha > \pi/6$, then we will conclude that $(|xr| + |qr|) - (|xp| + |pq|) < 0$ for any x and our proposition is proved.

Lemma 7. $|rp_{ro}| + |qr| - |pq| < 0$ for any r when $\alpha > \pi/6$.

Proof. We prove this by analyzing the function $h(\beta) = |rp_{ro}| + |qr| - |pq|$. We first show that the maximum of $h(\beta)$ is achieved when $\beta = \pi/2$, and then prove that for $\beta = \pi/2$ we have $|rp_{ro}| + |qr| - |pq| < 0$ when $\alpha > \pi/6$.

We first formulate $h(\beta)$: by defining

$$|qr| = \sin(\frac{\beta - \alpha}{2})$$
$$|rp_{ro}| = 1 - \cos(\pi - \beta - \alpha) = 1 + \cos(\alpha + \beta)$$
$$|pq| = 2\cos(\alpha)$$

we have

$$h(\beta) = |qr| + |rp_{ro}| - |pq| = 2\sin(\frac{\beta - \alpha}{2}) + 1 + \cos(\alpha + \beta) - 2\cos(\alpha).$$

We will show that $h(\beta)$ is a decreasing monotone function for $\pi/2 \leq \beta \leq \pi - \alpha$ (for $\alpha > \pi/6$) and, therefore, it reaches its maximum at $\beta = \pi/2$. We do this in the standard way, by differentiating h with respect to β and equating to 0. In

$$h'(\beta) = \cos(\frac{\beta - \alpha}{2}) - \sin(\alpha + \beta)$$

an extremum point is achieved when

$$\cos(\frac{\beta - \alpha}{2}) = \sin(\alpha + \beta).$$

This is achieved either for $\beta = (\pi - \alpha)/3$ or for $\beta = \pi - 3\alpha$. Neither of these cases occur for $\alpha > \pi/6$ and $\pi/2 \leq \beta \leq \pi - \alpha$. Thus, for $\alpha > \pi/6$, $h(\beta)$ is a monotone function for $\pi/2 \leq \beta \leq \pi - \alpha$. By comparing the end values we see that it is a decreasing monotone function and thus achieves its maximum at $\beta = \pi/2$.

All that is left to be done is prove that for $\beta = \pi/2$ (the value of β for which the maximum is achieved), $|rp_{ro}| + |qr| - |pq| < 0$ when $\alpha > \pi/6$. We do this by reformulating the sum for the special case $\beta = \pi/2$ (where $|qr| = |pr|$). Since the head angle of the triangle $\triangle prq$ is $(\pi + \alpha)$, we have

$$|rp_{ro}| = |qr| \cos(\pi/4 + \alpha/2),$$
$$|pq| \; = 2|qr| \sin(\pi/4 + \alpha/2).$$

Thus, $|rp_{ro}| + |qr| - |pq| = |qr|[\cos(\pi/4 + \alpha/2) + 1 - 2\sin(\pi/4 + \alpha/2)]$. Since $|qr|$ is positive and $\cos(\pi/4 + \alpha/2) + 1 - 2\sin(\pi/4 + \alpha/2)$ is a decreasing function of α, and is negative for $\alpha = \pi/6$, we have that $|rp_{ro}| + |qr| - |pq| < 0$ when $\alpha > \pi/6$ (for $\beta = \pi/2$ and hence for $\pi/2 \leq \beta \leq \pi - \alpha$). □

From Lemma 7 we therefore have that $|rp_{ro}| + |qr| - |pq| < 0$ for any r when $\alpha > \pi/6$. By Corollary 6 we have that $|rp_{rx}| \leq |rp_{ro}|$, and so

$$|rp_{rx}| + |qr| - |pq| \leq |rp_{ro}| + |qr| - |pq| < 0.$$

Since by Lemma 5 $(|xr| - |xp|) \leq |rp_{rx}|$, we easily see that

$$(|xr| - |xp|) + |qr| - |pq| \leq |rp_{rx}| + |qr| - |pq| \leq |rp_{ro}| + |qr| - |pq| < 0.$$

That is,

$$(|xr| + |qr|) - (|xp| + |pq|) < 0,$$

which terminates the proof. □

Voronoi Graph Matching for
Robot Localization and Mapping

Jan Oliver Wallgrün

University of Bremen,
Enrique-Schmidt-Str. 5, 28359 Bremen, Germany
wallgruen@informatik.uni-bremen.de

Abstract. In this article, we develop a localization and map building
approach for a mobile robot in which annotated generalized Voronoi
graphs are used as the robot's spatial representation of the environment.
The core of our approach is a matching scheme for solving the data association problem of identifying corresponding parts in two tree-formed
Voronoi graphs, one representing a local observation and the other one
representing the robot's internal map. Our approach is based on the
notion of edit distance which means it computes the cost optimal way
to transform both graphs into the same graph. The costs for adding or
deleting branches are based on a measure that assesses the stability of
nodes and edges in the graphs as well as their relevance for navigation. In
addition, we incorporate spatial constraints based on the graph annotations which leads to a top-down dynamic programming implementation
that performs reliably and efficiently in practice.

Keywords: Generalized Voronoi diagrams, Voronoi graphs, data association, matching, robot navigation, localization, map learning.

1 Introduction

Generalized versions of Voronoi diagrams [1,2] dealing with more complex geometric primitives (e.g., line segments [3,4]), instead of only point sites have
frequently been employed in the area of mobile robot navigation. For instance,
they have been considered as an intermediate representation for planning the mo-
tions of a robot [5,6]. Furthermore, they have more recently been employed as
a means to derive graph representations, so-called *topological maps* [7,8], which
serve as the robot's only spatial representation of its environment [9,10] (see
also Fig. 1). In contrast to more traditional representations such as occupancy
grids [11] or geometric representations [12], these representations are much more
compact and focus on the essential information required for path planning and
navigation. The main challenge with regard to this kind of representation approach is to develop robust techniques to learn and maintain the global graph
representation from local and uncertain sensor information gathered over time.
In robotics, this problem is generally referred to as the simultaneous localization
and mapping (SLAM) problem [13,14] and it is challenging because errors in the

M.L. Gavrilova et al. (Eds.): Trans. on Comput. Sci. IX, LNCS 6290, pp. 76–108, 2010.

localization and in map building affect each other and, hence, these errors can grow without bounds.

In this text, we consider Voronoi graph representations called *annotated generalized Voronoi graphs* which are derived from laser range data and focus on the *data association* or *correspondence* problem [15,16] of identifying corresponding parts in a local Voronoi graph extracted from the robot's current sensor measurements and in the (partially constructed) global Voronoi graph forming the robot's map. The data association problem is arguably one of the most fundamental and challenging subproblems underlying localization and map learning [17].

The data association algorithm we propose in this text is based on an inexact matching technique for annotated Voronoi graphs which takes into account that the underlying generalized Voronoi diagrams are very sensible to noise in the sensor data. For dealing with this problem we employ relevance and stability measures for Voronoi nodes and edges proposed in earlier work [18] to model the costs of different operations in our matching approach. The overall matching is then based on the edit distance paradigm for inexact graph matching [19,20]. In contrast to similar work on edit distance-based matching [21,22], in particular applications in the area of computer vision [23], our approach deals with graphs that only partially overlap and exploits additional information annotated to the graph structure tailored for the considered robot navigation scenario. For efficiency reasons, we restrict ourselves to matching tree-formed Voronoi graphs instead of complete graphs, enforcing acyclicity if needed. In addition to the edit distance-based matching, absolute position estimates and relative geometric constraints are applied to improve reliability and efficiency of the overall approach.

The remainder of the paper is structured as follows. We start by describing the underlying representation approach using annotated Voronoi graphs in Sect. 2. In Sect. 3, we introduce the relevance and stability measures together with an algorithm for their computation. In Sects. 4 and 5 we develop our Voronoi graph matching approach and in Sect. 6 we describe how the matching result can be used to update the map. The presented algorithms are then evaluated experimentally in Sect. 7.

2   Annotated Generalized Voronoi Graph Representation

The spatial representation we consider in this work are graphs derived from generalized Voronoi diagrams (GVDs) based on line segments as described in [3,4,1]. Given a polygonal description of an environment such as the one shown in Fig. 1(a), these GVDs consist of all points p of free space for which the maximal inscribed circle centered on p touches the obstacle boundaries at at least two different points. The points on the obstacle boundaries are called the *generating points* of p. If p has three or more generating points, it forms a *meet point* at which several Voronoi curves meet. The resulting GVDs are related to Blum's notion of a shape's skeleton [24]. In the derived graph representations, the nodes then stand for the meet and end points of the Voronoi curves in the GVD,

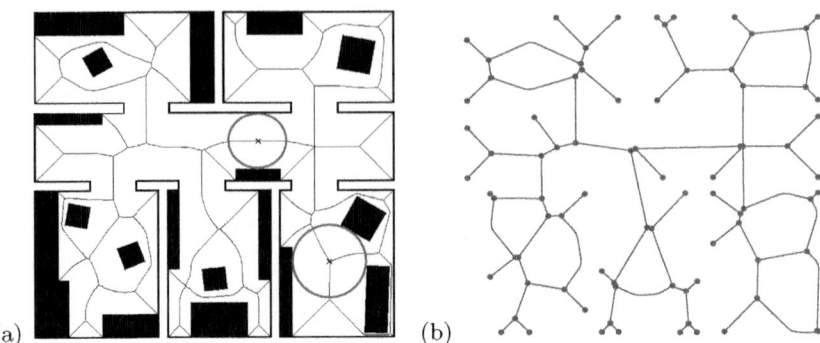

Fig. 1. (a) The generalized Voronoi diagram (fine lines) derived from a polygonal 2D environment including the maximal inscribed circles for two exemplary points, (b) the corresponding Voronoi graph representation

while the edges connect two nodes if the corresponding points are connected by a Voronoi curve (see Fig. 1(b)).

The graph structure is annotated with the following additional information:

1. Nodes are labeled with a *signature* (see Fig. 2(a)) that contains the distance to the generating points (which is the radius of the maximal inscribed circle) and the angles to the lines connecting the node with its generating points with respect to an arbitrarily chosen reference direction.
2. The approximate relative positions of nodes are represented by annotating the edges with the approximate distances between the connected nodes and by annotating the nodes with the approximate angles of the leaving edges, again with respect to the chosen reference direction (see Fig. 2(b)).
3. The combinatorial embedding of the graph into the plane is specified by storing the cyclic order of edges for each node.
4. Every edge is annotated with a description of the shape of the Voronoi curve corresponding to this edge. The Voronoi curve may deviate significantly from the direct connection between the two nodes. The description simply consists of a sequence of intermediate points in the local reference frame of the edge. We will adopt the term *course* [25] for this description. Moreover, the approximate travel distance for moving along the Voronoi curve is stored.
5. Additional edge attributes provide information about traversability of an edge (sufficient distance to obstacles) and whether the edge leads to still unexplored areas.

We will refer to the representation defined above as an *annotated generalized Voronoi graph* (AGVG). The signature information and relative position information will play a central role in our matching approach presented in Sect. 4. A benefit of using relative metric information is that the information can be globally inconsistent without diminishing its usability for navigation. The estimated travel distance annotated to the edges can be used for path planning by using one of the standard graph search techniques for weighted graphs.

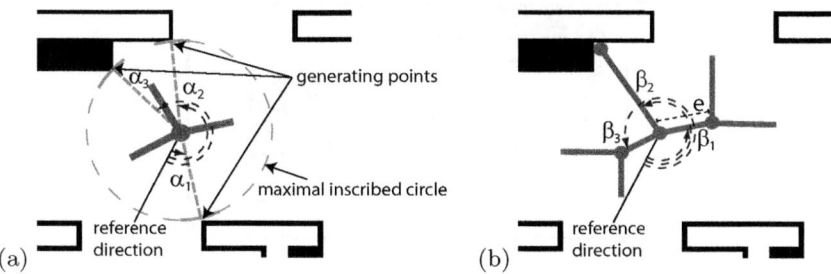

Fig. 2. Different kinds of annotations contained in an AGVG: (a) The node signature, (b) relative position information

3   Relevance and Stability Assessment

In [18] we defined relevance measures for the nodes and edges in an AGVG and demonstrated that these measures are well-suited to assess two things simultaneously: the entities' relevance for navigation and their stability under sensor noise. We also developed a simplification algorithm based on these measures and showed how it can be used to remove instable parts of an AGVG, to produce more abstract versions of it, or to build up a hierarchical version of an AGVG which describes the environment at different levels of granularity (see also [26] for details).

In contrast to similar work on shape representation and matching from the vision community (e.g., [27,28,29]) our techniques do not require a complete description of the shape of the boundaries of free space to simplify the Voronoi structures, but are purely based on the information stored in a (partial) AGVG. In the following we describe the relevance measures and how they can be computed.

3.1   Relevance Measures for Voronoi Nodes and Edges

For now we assume that we have a complete AGVG with no unexplored edges. For each inner node in an AGVG, the virtual lines connecting a node v with its generating points split up the free space of the environment into regions called *accessible regions* of v (see Fig. 3). We will use the notation $R_i^{[v]}$ for the region that can be accessed from node v via its i-th edge $e_i^{[v]}$ without crossing one of these virtual connections. If multiple leaving edges of a node are part of a cycle in the AGVG such as $e_2^{[A]}$ and $e_3^{[A]}$ in Fig. 3, their corresponding regions will be identical.

The significance value of an accessible region of a node v in an AGVG (which is based on the notions of distance and seclusion) is then given by the length of the path to the most distant node belonging to that region without counting that part of the path that lies inside the maximal inscribed circle of v. Assuming that the radius of the maximal inscribed circle of v is given by $r(v)$, $d(v_i, v_j)$ stands for the shortest path distance between nodes v_i and v_j in the AGVG,

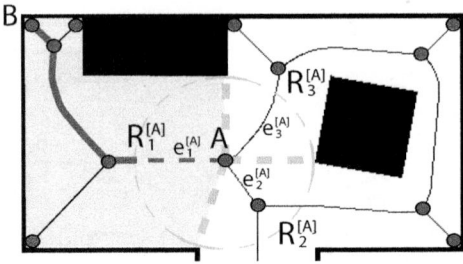

Fig. 3. Computation of the rsm value of region $R_1^{[A]}$ (shaded)

and the set $\mathcal{A}_i^{[v]}$ contains all nodes contained in $R_i^{[v]}$ (those nodes that can be reached from v via $e_i^{[v]}$ without passing v again), we define the *region significance measure* (rsm) as follows:

Definition 1 (Region significance measure). *The region significance measure rsm assigns a significance value from $\mathbb{R}_0^+$ to each accessible region $R_i^{[v]}$ of a node v which is given by:*

$$rsm(R_i^{[v]}) = \begin{cases} \infty & , \text{ if } e_i^{[v]} \text{ is part of a cycle} \\ \max\left\{ \left(\max_{w \in \mathcal{A}_i^{[v]}} d(v,w) \right) - r(v), 0 \right\}, \text{ else} \end{cases}$$

$$(1)$$

The first part of the definition assigns an rsm value of ∞ to accessible regions if the corresponding edge is part of a cycle. As a result, cyclic regions will be treated as maximally significant. Otherwise, the shortest paths to all the nodes belonging to the accessible region are considered to determine the length of the longest one. Finally, the radius of the maximal inscribed circle is subtracted from this path length. As a result, only the part of the path to the most distant node that lies outside the maximal inscribed circle is considered. Negative results for nodes located inside the maximal inscribed circle are turned into a relevance value of 0 by the outer maximum operation.

Figure 3 illustrates this approach: B is the most distant node of region $R_1^{[A]}$ (shaded region in the figure) in terms of shortest path length. The radius of the maximal inscribed circle is subtracted effectively removing the dashed part of the path so that only the solid drawn part is taken into account.

Based on this region significance measure, the overall relevance of a node v given by the introduced *Voronoi node relevance measure* (vnrm) is determined by taking the third highest rsm value over the rsm values of all accessible regions of v because a node can only be considered a relevant decision point for navigation if it has at least three significant accessible regions:

Definition 2 (Voronoi node relevance measure). *The Voronoi node relevance measure vnrm : $V \to \mathbb{R}_0^+$ assigns each Voronoi node v with $\deg(v) \geq 3$ a relevance value given by*

$$vnrm(v) = \max{}^3 RSM_v$$

$$(2)$$

where RSM_v is the multiset $\{rsm(R_i^{[v]}) \mid 1 \le i \le \deg(v)\}$ and $\max^3$ stands for the third highest element.

As a result of this definition, a node which has three or more edges which are part of cycles in the AGVG will be assigned a vnrm value of ∞. In Fig. 4(a), the individual vnrm values of the nodes in our exemplary environment are visualized by the radii of the corresponding circles. Nodes that have a vnrm value of ∞ are displayed by the non-filled circles. Important decision points, such as the junctions on the central corridor, receive high values while nodes caused by noise or by minor concavities in the obstacle boundaries have very low vnrm values. By cutting off branches in order of increasing rsm values and up to a given threshold, simplified versions of the original AGVG can be derived such as the one shown in Fig. 4(b).

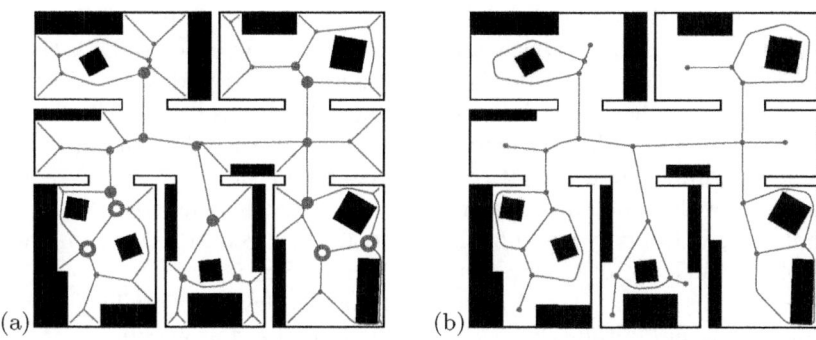

Fig. 4. (a) vnrm values assigned to the nodes of an AGVG depicted by the radii of the corresponding circles (non-filled circles stand for nodes with a vnrm value of ∞), (b) a simplified AGVG derived by cutting off irrelevant branches up to a certain threshold

3.2 Admitting Incomplete Information

In the context of mobile robot mapping, most of the time no complete AGVG is available as we assumed in the previous sections. Either the AGVG is a local graph derived from sensor information about the robot's immediate surroundings or it is a partially constructed map AGVG built up by merging local AGVGs. In both cases, the AGVG contains edges that are marked as unexplored. We deal with this by first treating all nodes of degree one that mark the end of a still unexplored edge in the same way as the corner nodes and compute the relevance values as defined above. Without knowing the complete AGVG, the computed values will often only be lower bound estimates of the actual relevance values: Every rsm value computed for a non-cyclic region in which the local AGVG has unexplored edges will be a lower bound of the real rsm value and we will mark it as such by introducing a new boolean attribute $lb(R_i^{[v]})$.

For the vnrm values we proceed similarly by recording that the vnrm value of a node v is only a lower bound estimate (node attribute $\mathrm{lb}(v)$ is true) if the third most relevant region or an even less relevant region of the node has a rsm value which is a lower bound estimate. The existence of lower bound estimates of nodes in a partially constructed map AGVG means that these need to be updated whenever new parts are added to the AGVG.

3.3   Computing the Relevance Values

In the following, we present an algorithm for actually computing the rsm values for the accessible regions of Voronoi nodes in an AGVG and, based on the result, the vnrm values of the nodes themselves. We assume that we have a complete AGVG at hand. Extending the approach to deal with partial AGVGs by setting the respective node attributes is straightforward.

We start by describing the computation of the rsm values of the accessible regions of a single node v. Our algorithm is a modified version of Dijkstra's single source shortest path algorithm [30] which determines the distance from a given start node to all other nodes in a weighted graph. Naturally, v is the source node and the length of the Voronoi curves annotated to the edges of the AGVG are taken as weights. The modifications we make are for detecting cyclic regions and for tracking which node has been reached via which leaving edge of the starting node v. A pseudocode version of the algorithm is given in Algorithm 1 and will be explained below. An example run is given in Fig. 5.

The algorithm uses a set of auxiliary variables that for simplicity are treated as global variables, although they are actually realized as attributes of the node and edge objects. These variables are

- rsm_i (initialized to 0), which contain the current estimates of the rsm values of the accessible regions $R_i^{[v]}$ and, thus, in the end contain the result of the algorithm,
- cyc_i (initialized to false), which are set to true when edge $e_i^{[v]}$ is detected to be part of a cycle,
- d_{v_i}, storing the current shortest distance of v_i from v,
- m_{v_i} (initialized to 0), which are used to mark nodes based on which accessible region of v they belong to ($\mathrm{m}_{v_i} = 0$ means that this node has not been reached yet),
- closed_{v_i} (initialized to false), which, as in the standard Dijkstra algorithm, mark whether a node has been closed, meaning its distance from v has been determined,
- a set L, which contains the nodes that have been reached and still need to be expanded.

Initialization. In the initialization step of the algorithm the auxiliary variables are initialized as stated above. L is initially empty. For v, m_v is set to -1, which is a special mark only used for the start node, d_v is set to 0, and v is marked as closed. Then a single expansion step is performed for v by calling the

---

Algorithm 1. Relevance computation algorithm for a Voronoi node v

---

procedure computeRelevanceValues(Node v)

1: $rsm_i \leftarrow 0$, for all $1 \leq i \leq \deg(v)$ {initialization}
2: $cyc_i \leftarrow$ false, for all $1 \leq i \leq \deg(v)$
3: $closed_{v_i} \leftarrow$ false, for all $1 \leq i \leq |V|$
4: $m_{v_i} \leftarrow 0$, for all $1 \leq i \leq |V|$
5: $L \leftarrow \emptyset$
6: $m_v \leftarrow -1$; $d_v \leftarrow 0$; $closed_v \leftarrow$ true
7: **for all** $e_i^{[v]}, 1 \leq i \leq \deg(v)$ **do**
8:     expand($e_i^{[v]}$,v,i)
9: **end for**
10: **while** $|L| > 0$ **do** {main loop}
11:     $w \leftarrow \text{argmin}_{l \in L}\, d_l$
12:     $L \leftarrow L \setminus w$
13:     **if** not $closed_{m_w}$ and not cyc_{m_w} **then**
14:         $closed_w \leftarrow$ true
15:         $rsm_{m_w} = \max\{d_w - r(v), 0\}$
16:         **for all** $i, 1 \leq i \leq \deg(w)$ **do**
17:             expand($e_i^{[w]}$,w,m_w)
18:         **end for**
19:     **end if**
20: **end while**
21: $vnrm_v \leftarrow \max^3\{rsm_i \mid 1 \leq i \leq \deg(v)\}$

procedure expand(Edge e, Node x, Integer i)

1: $y \leftarrow \text{other}(e, x)$
2: **if** not $closed_y$ **then**
3:     **if** $m_y \neq 0$ and $i \neq m_y$ **then**
4:         $cyc_i \leftarrow$ true; $rsm_i = \infty$
5:         $cyc_{m_y} \leftarrow$ true; $rsm_{m_y} = \infty$
6:     **else if** $m_y = 0$ or $d_x + l(e) < d_y$ **then**
7:         $d_y \leftarrow d_x + l(e)$; $m_y \leftarrow i$
8:         $L \leftarrow L \cup \{y\}$
9:     **end if**
10: **else if** $m_x = -1$ **then**
11:     $cyc_i \leftarrow$ true; $rsm_i \leftarrow \infty$
12: **end if**

---

subprocedure expand (to be explained in detail below) for each leaving edge of v. This expansion puts each node adjacent to v into the list L and marks it with the number i of the edge $e_i^{[v]}$ by which it has been reached. In addition, the distance values d_{v_i} are updated according to the length of the edge. The region numbers will be propagated through the graph, in addition to the minimal distances being computed in the main part of the algorithm. Figure 5(b) illustrates the state after the initialization step: v has been expanded. Its neighbors are now contained in L (depicted by the surrounding circles) and are marked with the region numbers 1 to 3. All nodes without a number still are marked as 0. The rsm_i variables for v are still 0.

Main Loop. The main loop operates similarly to the standard Dijkstra algorithm: Node w with the current minimal distance d_w is taken from L and expanded. Neighboring nodes are labeled with the same region number as w and put into L. In addition, as the minimal distance of w has hereby been determined, we update the rsm_i variable for the corresponding region of v according to Eq. 1. Figure 5(c) and (d) show this step for nodes A and B, respectively, resulting in rsm values of 5 and 7 based on the length of the edges. After expanding B, L now contains five nodes: two for region $R_1^{[v]}$, one for $R_2^{[v]}$, and two for $R_3^{[v]}$.

Expand Procedure. The expand procedure performs the expansion of one edge connecting the currently considered node x with another node y. There are three main cases that need to be distinguished:

1. $m_y = 0$: This means node y has not been reached yet. In this case y will be marked as belonging to the same region as x (m_y is set to i); its distance d_y will be set to d_x plus the length $l(e)$ of the connecting edge e. Finally, y will be added to L.
2. $m_y > 0$ and $i = m_y$: y has already been reached and is labeled with the same region number as the one propagated from x. If the distance d_x plus the length of the connecting edge is smaller then the current distance d_y, we have discovered a new shortest path to y, and d_y is updated.
3. $m_y > 0$ and $i \neq m_y$: Again, y has already been reached, but this time is labeled with a different region number than the propagated one. This means a cycle involving two leaving edges of v has been detected. Accordingly, the cyc_i flags are set for both involved edges of v, signaling that no further expansion is required for these regions. Their rsm_i variables are set to ∞.

We have already seen the first case applied in Figs. 5(b)–(d). In Fig. 5(e), C is expanded and a cycle is detected (case 3) because the adjacent node D is labeled with the region number 1, not 2. As a result, rsm_1 and rsm_2 are set to ∞ and expansion stops in this part of the graph.

In Fig. 5(f), E is expanded, which leads to an updated rsm variable for region 3 because we have determined the shortest path to a new most distant node. Here, we encounter an instance of case 2 because E's neighbor F is labeled with the same region number. As the distance via E is larger than the already known distance for E, no distance update takes place.

Figs. 5(g)–(i) show three more expansion steps in region 3. In the last two expansions, no more nodes are added to L; only rsm_3 is updated. Eventually, L is empty and the rsm_i variables contain the rsm values of all three regions. The third highest value (in this case 17) is the vnrm value of v.

In the actual implementation of the algorithm the list L is realized as a priority queue sorted by increasing distance d_{v_i}. As for the standard Dijkstra algorithm, the time complexity is $O(|E| + |V| \log |V|)$ because of the need to access nodes in order of increasing distance from the start node. The fact that an expansion is stopped when a region has been shown to be cyclic improves the efficiency in practice, especially in graphs with many cycles.

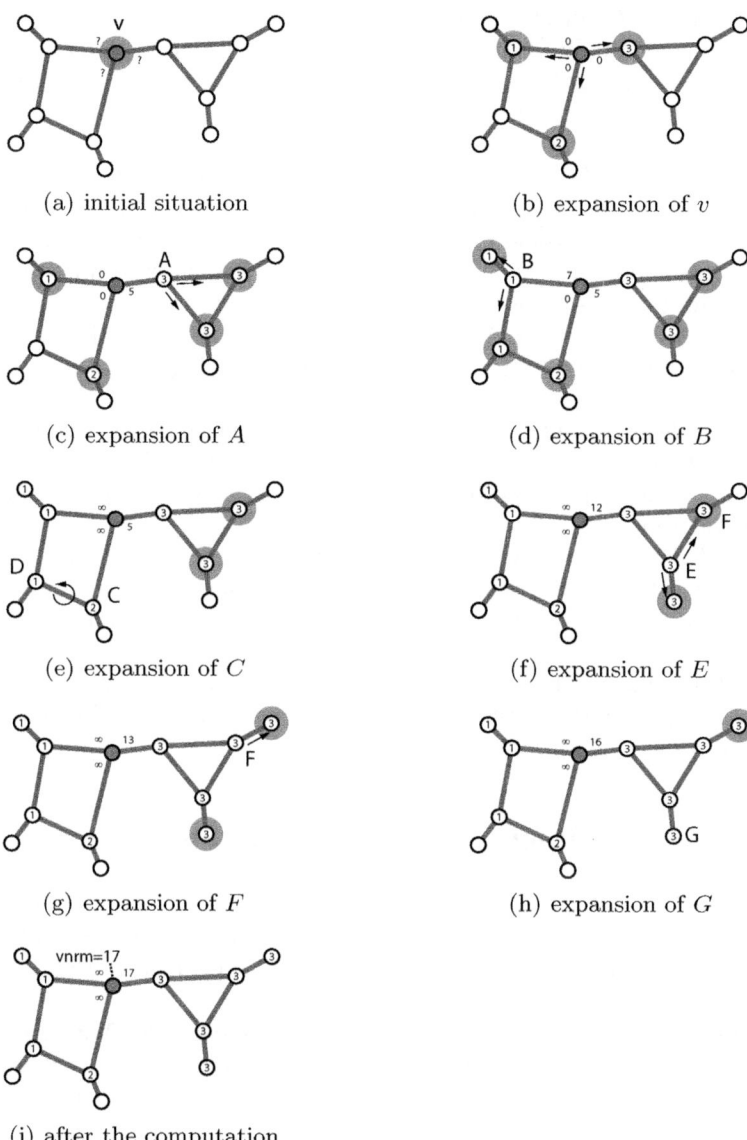

(a) initial situation

(b) expansion of v

(c) expansion of A

(d) expansion of B

(e) expansion of C

(f) expansion of E

(g) expansion of F

(h) expansion of G

(i) after the computation

Fig. 5. Relevance value computation for node v performed by our algorithm

3.4   Complexity and Optimizations

Computation of the rsm and vnrm values for all nodes in the graph changes the problem into an all-pair shortest path problem in which the shortest paths from every node to every other node have to be determined. No algorithm for this problem is known that has a better worst-case complexity than what we get when applying Dijkstra's single source shortest path algorithm for every node independently. This results in an $O(|V|^2 \log |V| + |E||V|)$ time complexity. Accordingly, we execute our relevance computation method for each node with degree 3 or higher. The computed rsm and vnrm values are stored as annotations to the nodes and edges in the graph for later use.

To increase the efficiency of the basic relevance computation algorithm described above, we made several fundamental optimizations. We realized that determining the vnrm value of a node only requires the precise rsm values for the third most relevant regions and the regions even less relevant than this. In addition, the two most relevant regions do not play a role when it comes to pruning the AGVG by removing the unstable parts. Based on these insights, we optimized the relevance computation algorithm by doing the following:

- As simplification and computation of vnrm values does not require the rsm values of the two most relevant regions for a given node, rsm value computation can be stopped as soon as it becomes clear which two regions are the most relevant ones.
- When AGVGs are constructed incrementally and the current AGVG is complemented by a new subgraph, it is not necessary to recompute relevance values for nodes for which only the two most relevant regions have changed.

As a result of improvements made based on these ideas, the relevance computation becomes very fast in practice. A quantitative analysis of this improvement is given in Sect. 7.1.

4   AGVG Matching for Data Association

As mentioned, one of the main problems in robot navigation and mapping is the data association or correspondence problem of identifying which features from the local perception (the *data set*) correspond to which feature from the robot's internal spatial model (the *model set*).

For AGVGs, it makes sense to assume that no two perceived Voronoi nodes can correspond to the same one in the model set. Still, the number of possible data associations is huge. Hence, additional information needs to be exploited in order to come up with efficient and reliable solutions to the data association problem. Examples of information that has been employed for this purpose are feature similarity, position estimates, co-visibility, and configurational knowledge.

Most work on the data association problem in robotics deals with point landmarks or simple geometric features and often very simple methods like ICNN [31] (basically validation gating [15] applied to the Mahalanobis distance [32] between data and model feature) are used in which only the individual compability of pairs of features from the data and model set are assessed. Overall, these approaches are rather fast but often not very reliable.

To improve the reliability one has to consider the joint compability of the associations for the complete set of simultaneously observed features, an approach often called *batch association* [31,33]. Several batch association approaches explicitly employ hard geometric constraints based on the configuration of features to filter out data associations which are not jointly compatible [34,35,33,36].

In the following, we develop a batch association approach for AGVGs which is based on graph matching techniques exploiting the fact that graph topology and combinatorial embedding restrict the set of valid data associations. Our approach is based on inexact graph matching techniques because we need to take into account variations resulting from the instability of the underlying GVD. A typical situation showing two consecutively perceived AGVGs overlayed based on an estimate of the robot pose is shown in Fig. 6. Both AGVGs contain branches that are not contained in the other one.

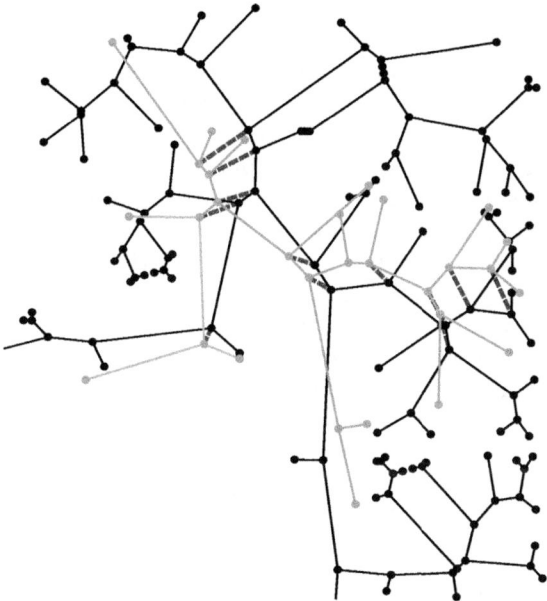

Fig. 6. A typical AGVG matching problem

We restrict ourselves to the matching of tree-formed AGVGs because tree matching problems can in most cases be solved more efficiently than the corresponding graph matching problems[1] and it does not significantly restrict the applicability of the approach. In practice, we enforce acyclicity in the AGVGs by splitting the most distant edges that belong to a cycle.

Since our AGVGs are combinatorially embedded, our matching is one between ordered trees. This is a crucial property for our approach because computing the edit distance for unordered labeled trees has been shown to be NP-complete

---

[1] For instance, tree isomorphism and subtree isomorphism problems can be solved in polynomial time [37].

[22], while it is in P for ordered trees [21]. We start the presentation of our data association algorithm by first defining the underlying tree matching together with the required edit operations and their costs. We then extend this approach by incorporating additional constraints in the next section.

4.1   Ordered Tree Matching Based on Edit Distance

The basic idea of the edit distance approach for inexact tree (or graph) matching is to define costs for operations that modify the trees (e.g., adding or deleting nodes and edges) and compute a cost-minimal sequence of operations that transforms both trees into the same tree. This sequence of transformation operations directly implies a matching between the nodes (and edges) of the graphs.

In our approach to matching two tree-formed AGVGs G and G', we start by considering all possible initial matches of a node u from G with a node u' from G'. This choice turns the AGVGs into *rooted trees* with respective root nodes u and u'. In addition to being rooted, the fact that for each node the edges are cyclically ordered as specified by the AGVGs' combinatorial embedding means that the resulting trees also are *ordered*. Hence, if u and u' have the same degree n, there are n ways of mapping the edges of u to those of u' while preserving the cyclic orders, a process that we will refer to as *aligning*. The different alignment variants can then be specified by providing a single edge offset parameter. For each start matching $u \rightsquigarrow u'$ and edge offset i, the edit distance can be computed from the edit distances of matching the corresponding subtrees of u and u'. Thus, the key component of our approach is an edit distance $\mathrm{dist}_{\mathrm{st}}$ for subtrees in an AGVG and we start by defining the required edit operations, their costs, and how the subtree edit distance is computed.

4.2   Edit Distance for Subtrees in Two AGVGs

In the remainder of this text, we will use the following notations: For nodes of the observed data AGVG G we use u, v, w, etc. while using u', v', w', etc. for nodes from the model or map AGVG G'. We will denote a subtree with root v and ancestor u as $\triangle_v^{(u)}$. Providing the ancestor u here only serves the purpose of specifying which of v's edges leads upward in the rooted tree, namely the one leading to u, and which ones lead to v's child nodes. Hence, we put the ancestor label in brackets.

Given two subtrees and assuming that their roots correspond, the correspondences between their edges are uniquely determined. We write $\mathrm{child}_i(v)$ for the i-th child node of node v in a given subtree which is the node connected to the i-th successor edge of the edge connecting v to its parent in the cyclic edge order.

For the root node u of a rooted AGVG, only the order of child nodes is determined, but it is not clear which adjacent node should be considered as the first one. For such cases we introduce the function $\mathrm{neighbor}_i(v)$ to refer to the node connected to v via edge $e_i^{[v]}$. In addition, we will use the notation $i \oplus j$ for the index of the j-th successor of the element with index i in a cyclically ordered set. For instance, $\text{child}_{(2\oplus2)}(v)$ for a node with three children is equivalent to $\text{child}_1(v)$. $\text{neighbor}_{(1\oplus2)}(v)$ with $\deg(v) = 3$ is equivalent to $\text{neighbor}_3(v)$.

Fig. 7 shows an example situation of two subtrees $\triangle_v^{(u)}$ and $\triangle_{v'}^{(u')}$ with roots v and v' that should be compared. We consider three different matching cases which directly correspond to our edit operations. The names of the operations are chosen from the perspective of modifications made to the subtree of the local AGVG.

1. **Match operation:** The match operation implies that the two root nodes v and v' will be considered as matched. This means that no modification is required, but the costs for matching each of the subtrees formed by the child nodes of v to those of the corresponding child nodes of v' have to be considered. We assume here that v and v' have the same degree and, hence, the same number of child nodes. As we will see later, we will reject matchings for which this is not the case beforehand (cf. Sect. 5.1).

2. **Delete operation:** With this operation we conclude that node v is not contained in the model AGVG and, hence, it has to be deleted from subtree $\triangle_v^{(u)}$ together with all the subtrees formed by its child nodes except one. This remaining child subtree will then need to be matched to the model subtree $\triangle_{v'}^{(u')}$. Consequentially, the delete operation actually subsumes $\deg(v) - 1$ different cases of modification, one for each subtree of v. To compute the minimal costs, all cases have to be considered and the one that results in the lowest costs has to be chosen.

3. **Add operation:** The add operation is symmetrical to the delete operation. We assume here that the root node of the model subtree has not been perceived so that this node would have to be added to the local AGVG before node v. Analogously to the delete case, now one of the child subtrees of v' needs to correspond to $\triangle_v^{(u)}$, while all other subtrees would have to be added together with v'. Again, all cases have to be considered to determine the one that results in minimal costs.

Based on the three operations described above, we can now recursively define the edit distance of two rooted and ordered subtrees $\triangle_v^{(u)}$ and $\triangle_{v'}^{(u')}$. First, the edit distance $\text{dist}_{\text{st}}(\triangle_v^{(u)}, \triangle_{v'}^{(u')})$ is the minimum over the costs resulting from applying the individual operations: $\text{dist}_{\text{m}}(\triangle_v^{(u)}, \triangle_{v'}^{(u')})$ for the costs of the match operation, $\text{dist}_{\text{d}}(\triangle_v^{(u)}, \triangle_{v'}^{(u')})$ for the delete operation, and $\text{dist}_{\text{a}}(\triangle_v^{(u)}, \triangle_{v'}^{(u')})$ for the add operation. In addition, the recursion terminates if either v or v' marks the end of a still not completely explored edge. We introduce the node attribute $\text{uk}(v)$ to indicate if this is the case. If $\text{uk}(v)$ is true, the costs for matching the subtrees are simply 0. If $\text{uk}(v')$ is true, we want the costs to be based on $\triangle_v^{(u)}$ in order to penalize leaving parts of the local AGVG unmatched. For now, we only introduce a cost function $\text{unmatched}(\triangle_v^{(u)})$. This function will be defined later in this section. The overall dist_{st} edit distance then is given by:

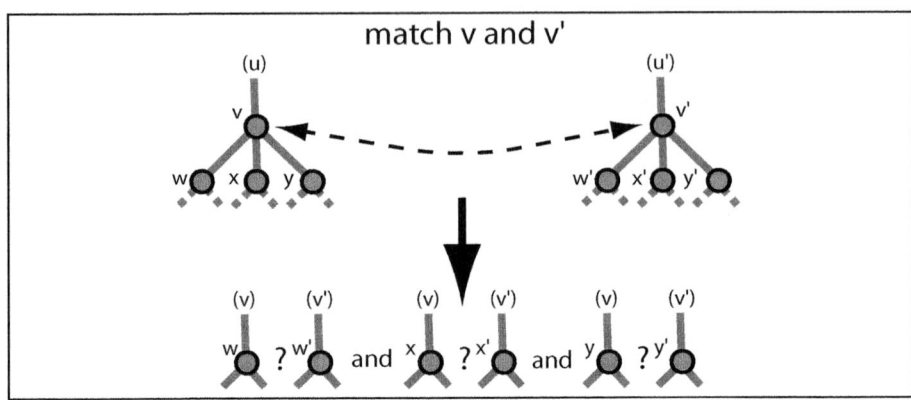

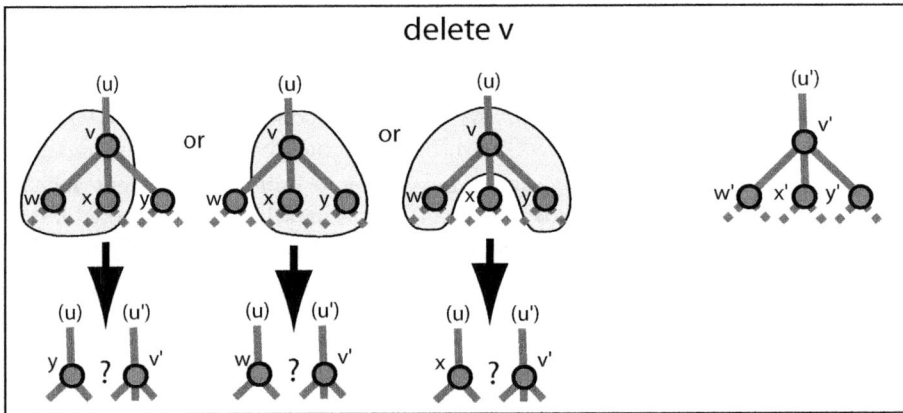

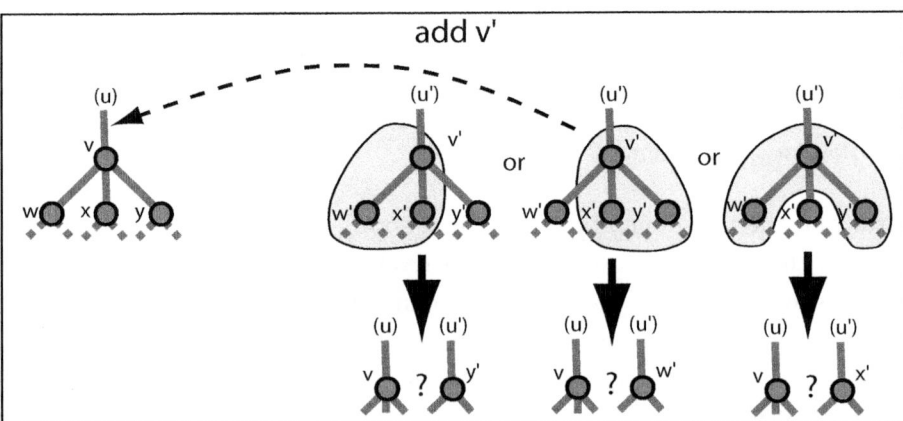

Fig. 7. The three edit operations for matching subtrees from two AGVGs: matching the root nodes, deleting the root of the local AGVG, and adding the root of the model AGVG

$$\text{dist}_{\text{st}}(\triangle_v^{(u)}, \triangle_{v'}^{(u')}) =$$

$$\begin{cases} 0 & , \text{if uk}(v) \\ \text{unmatched}(\triangle_v^{(u)}) & , \text{if uk}(v') \\ \min\left\{\text{dist}_{\text{m}}(\triangle_v^{(u)}, \triangle_{v'}^{(u')}), \text{dist}_{\text{d}}(\triangle_v^{(u)}, \triangle_{v'}^{(u')}), \text{dist}_{\text{a}}(\triangle_v^{(u)}, \triangle_{v'}^{(u')})\right\} & , \text{else} \end{cases} \quad (3)$$

The edit distance $\text{dist}_{\text{m}}(\triangle_v^{(u)}, \triangle_{v'}^{(u')})$ resulting from applying the match operation is the sum of the edit distances resulting from matching corresponding child subtrees:

$$\text{dist}_{\text{m}}(\triangle_v^{(u)}, \triangle_{v'}^{(u')}) = \sum_{i=1}^{\deg(v)-1} \text{dist}_{\text{st}}\left(\triangle_{\text{child}_i(v)}^{(v)}, \triangle_{\text{child}_i(v')}^{(v')}\right) \quad (4)$$

The edit distance $\text{dist}_{\text{d}}(\triangle_v^{(u)}, \triangle_{v'}^{(u')})$ for the delete operation is defined based on two components: First, $\text{rem}_j(v)$ are the costs for removing the subtree corresponding to the j-th child of v. We will later define these costs based on the rsm values of the corresponding accessible region introduced in Sect. 3. Second, we have to add the costs of matching one of the child subtrees to $\triangle_{v'}^{(u')}$ as given by dist_{st}. To determine the best case, we have to consider all $\deg(v) - 1$ possibilities and for each possibility sum up the costs for removing the other child subtrees. Then we add the costs for the recursive matching to it. Deletion is not possible if $\deg(v) = 1$ because we have arrived at an end node where a Voronoi curve ends in a corner of the environment. In this case the costs for the delete operation will be ∞. $\text{dist}_{\text{d}}(\triangle_v^{(u)}, \triangle_{v'}^{(u')})$ then is defined as follows:

$$\text{dist}_{\text{d}}(\triangle_v^{(u)}, \triangle_{v'}^{(u')}) =$$

$$\begin{cases} \min_{1 \le i \le \deg(v)-1} \left\{ \underbrace{\left(\sum_{j=1, j \ne i}^{\deg(v)-1} \text{rem}_j(v)\right)}_{\text{removing all except } i\text{-th subtree}} + \underbrace{\text{dist}_{\text{st}}\left(\triangle_{\text{child}_i(v)}^{(v)}, \triangle_{v'}^{(u')}\right)}_{\text{matching } i\text{-th subtree}} \right\} & , \deg(v) > 1 \\ \infty & , \text{else} \end{cases} \quad (5)$$

The costs for the add operation are computed accordingly, this time employing $\text{add}_j(v')$ for the costs of adding the entire j-th child subtree of v' to $\triangle_v^{(u)}$. Similarly to the delete case, the costs will be ∞ if $\deg(v') = 1$ as no terminal node can be inserted in the middle of an edge.

$$\text{dist}_{\text{a}}(\triangle_v^{(u)}, \triangle_{v'}^{(u')}) =$$

$$\begin{cases} \min_{1 \le i \le \deg(v')-1} \left\{ \underbrace{\left(\sum_{j=1, j \ne i}^{\deg(v')-1} \text{add}_j(v')\right)}_{\text{adding all except } i\text{-th subtree}} + \underbrace{\text{dist}_{\text{st}}\left(\triangle_v^{(u)}, \triangle_{\text{child}_i(v')}^{(v')}\right)}_{\text{matching } i\text{-th subtree}} \right\} & , \deg(v') > 1 \\ \infty & , \text{else} \end{cases} \quad (6)$$

Clearly, when computing the minimal distance for different choices of start nodes, we have a problem with overlapping subproblems (the computation of dist_{st} for a particular subtree) and optimal structure (dist_{st} is computed recursively from the edit distances of smaller subtrees). Hence, we employ a dynamic programming approach to compute the overall best matching. The already computed dist_{st} values for pairs of subtrees from the observed AGVG and the model AGVG are stored either in an array or in a hash table. As each edge in an AGVG defines two subtrees, one for each orientation of the edge, the maximal number of entries is $2 \times e \times 2 \times f$ where e is the number of edges in the observed AGVG and f in the model AGVG. In addition to the costs, we store the operation that resulted in the lowest costs and in case of the delete or add operation we also store the parameter i which resulted in the minimal edit distance. This allows for reconstructing the sequence of operations that lead to the minimal edit distance.

We now proceed by explaining how the edit distance of subtrees is used to compute the best overall matching.

4.3   Overall Edit Distance

Based on the edit distance for subtrees, we can now define $\text{dist}_{\text{node}}(u, u')$, the edit distance for an initial matching of two nodes u and u'. The overall edit distance of two AGVGs $\text{dist}_{\text{AGVG}}(G, G')$ is then defined as the optimum over all possible initial matchings.

As we already mentioned, when considering a start matching of two compatible nodes u and u', both of degree n, there exist n possible ways of aligning the edges of u and u' in a way that preserves the cyclic order information. Assuming that the minimal cost for the variant with edge offset i is given by $\text{dist}_{\text{aligned}}(u, u', i)$, the costs for this start matching are given by the minimum over all possible i:

$$\text{dist}_{\text{node}}(u, u') = \min_{1 \leq i \leq \deg(u)} \text{dist}_{\text{aligned}}(u, u', i) \tag{7}$$

For an alignment given by edge offset i, the edit distance $\text{dist}_{\text{aligned}}(u, u')$ is given by the sum of the costs of transforming the subtrees of corresponding edges into each other:

$$\text{dist}_{\text{aligned}}(u, u', i) = \sum_{j=1}^{\deg(u)} \text{dist}_{\text{st}} \left(\triangle^{(u)}_{\text{neighbor}_j(u)}, \triangle^{(u')}_{\text{neighbor}_{(j \oplus i)}(u')} \right) \tag{8}$$

The overall edit distance is then computed as the minimum over all possible start matchings:

$$\text{dist}_{\text{AGVG}}(G, G') = \min_{v \in V(G), v' \in V(G')} \left\{ \text{dist}_{\text{node}}(v, v') \right\} \tag{9}$$

The set of recursive equations Eq. 3–9 directly describes a first basic version of our matching algorithm. Following the dynamic programming approach, every time we require a $\mathrm{dist_{st}}$ value, we first check if this value has already been computed and, thus, is stored in the table. Only if this is not the case, the recursive computation is triggered and the result is then added to the table.

In addition to restricting the computation of $\mathrm{dist_{st}}$ values to those that are really required for the AGVGs at hand, we utilize branch and bound techniques at the global level (by comparing the current distance to the currently best solution) and the locally (when computing the best operation) to further increase the efficiency of the edit distance computation.

4.4   Modeling Removal and Addition Costs

As our edit distance approach is supposed to model the typical variations of AGVGs resulting from sensor limitations, the relevance and stability measures developed in Sect. 3 are well suited to describe the costs for deleting or adding subtrees. Therefore, the costs for deleting or adding subtrees of a node u, $\mathrm{add}_j(u)$ and $\mathrm{rem}_j(u)$, are defined as the rsm values of the corresponding accessible regions $R_j^{[u]}$:

$$\mathrm{add}_j(u) = \mathrm{rem}_j(u) = \mathrm{rsm}(R_j^{[u]}) \tag{10}$$

As we store the rsm values as attributes to the edges, we do not need to consider the other nodes contained in the subtrees that are supposed to be removed. Hence, computing the complete sums in Eq. 5 and Eq. 6 only requires $O(k)$ time with $k = \deg(v)$.

The costs $\mathrm{unmatched}(\triangle_v^{(u)})$ for leaving a complete subtree unmatched when the other subtree consists of an unknown node are determined in the same way: We simply add up the rsm values of the accessible regions $R_i^{[v]}$ which correspond to the i-th child of v.

$$\mathrm{unmatched}(\triangle_v^{(u)}) = \sum_{i=1}^{\deg(v)-1} \mathrm{rsm}(R_i^{[v]}) \tag{11}$$

As a result of choosing the penalty costs for unmatched parts to be of the same order of magnitude as the costs for removal and addition operations, we avoid that the matching algorithm prefers solutions with only very little overlap between the two AGVGs. This approach is advantageous in our case because for applications such as incremental mapping or tracking of Voronoi nodes the overlap between compared AGVGs is generally very large. For other kinds of applications, a different approach for balancing the individual costs might be better suited.

4.5   Complexity

The exploitation of the ordered tree structure in the matching process results in a data association algorithm with a rather low time complexity considering that the general interpretation tree is exponentially sized. As mentioned above, the size of a table storing all dist_{st} values is $4 \times e \times f$ for a local AGVG with e edges and a model AGVG with f edges. As we consider trees, this lies in $O(mn)$ where m and n are the number of nodes in the trees, respectively. Computing a table entry when the entries of the required subtrees are already known, requires $O(k^2)$ time where k is the maximal node degree occurring in both trees (which is typically 3). The quadratic dependency on k results from the computation of dist_{d} and dist_{a} (see Eq. 5 and 6). Hence, setting up the complete table can be done in $O(mnk^2)$ time.

The computation of $\text{dist}_{\text{node}}$ from a set up table takes $O(k^2)$ time because k table entries have to be added for k ways of aligning the edges of the two nodes. Finally, the overall $\text{dist}_{\text{AGVG}}$ function for matching two trees requires $m \times n$ times the computation of $\text{dist}_{\text{node}}$. As a result, the overall time complexity of the matching remains in $O(mnk^2)$.

As k can be considered a small constant for AGVGs, the resulting $O(mn)$ time complexity of the matching algorithm is of the same order as of the standard nearest neighbor (ICNN) algorithm. In addition, it only yields data associations that comply to the constraints arising from graph topology and combinatorial embedding. Moreover, because of the top-down approach and optimizations described in the previous section, it is usually sufficient to compute a very small fraction of all table entries. The effects of these additional efficiency improvements are hard to grasp theoretically and, hence, are not reflected in the coarse complexity assessment given here. As we will see in the next section, a further reduction of matchings that actually need to be considered can be achieved by incorporating hard constraints based on additionally available information.

5   Incorporating Constraints into the Matching

In the following, we extend the basic edit distance approach for matching tree-formed AGVGs by incorporating a set of hard constraints into the matching process. This exploitation of additional information further reduces the number of valid matchings, but the extensions we have to make in order to incorporate certain kinds of constraints increase the worst-case complexity of the matching approach. In exchange, we gain a significant improvement of the matching quality. The restriction to local constraints holding between neighboring entities ensures that the computational costs remain acceptable. In principle, it would have been possible to model the restrictions stemming from the additional information as soft constraints and to incorporate them as additional costs into the edit distance. However, we prefer hard constraints as they allow to reject matchings early and no additional parameters for weighing the constraints are required.

We consider the following constraints which are classified based on their arity:

1. unary constraints:
   (a) matching node degrees
   (b) absolute distance constraints for Voronoi nodes based on estimated robot pose

(c) absolute distance constraints for corresponding generating points
(d) unary node compatibility constraints based on the node attributes
2. binary relative distance constraints holding between adjacent nodes (in terms of both, line-of-sight distance and distance based on edge length)
3. ternary constraints about the relative angles formed at the nodes by the incident edges.

5.1 Unary Constraints

Unary hard constraints concern the compatibility of two features that are supposed to be matched with each other. The first condition we already mentioned previously is that the node degrees of two nodes v and v' need to agree in order to be matched:

$$\text{compatible}_{\text{deg}}(v, v') \iff \deg(v) = \deg(v') \tag{12}$$

When absolute position estimates for the nodes are available, these can be used to restrict possible matchings in the same way this is done in ICNN using validation gating (we here assume that position estimates are stored as additional node attributes using Gaussian probability distributions). Hence, another unary constraint we consider is that the Mahalanobis distance $M^2(v, v')$ between two nodes v and v' is below a given threshold γ_M[2]:

$$\text{compatible}_{\text{absdist}}(v, v') \iff M^2(v, v') < \gamma_M \tag{13}$$

Similarly, whenever absolute position information is available, we can define a criterion $\text{compatible}_{\text{absdist-gp}}(v, v')$ that requires that the Mahalanobis distances between every pair of corresponding generating points of v and v' is also smaller than γ_M.

Moreover, we can capture the compatibility of two nodes by looking at the local geometry as described by their signatures. This information consists of the radii $r(v)$ of the nodes' respective maximal inscribed circle and angles between the connections to the generating points. Fig. 8 illustrates how we measure the relative angles for two nodes with degree three: As the alignment between the two nodes is given by the edge offset parameter i, generating point $gp_j^{[v]}$ of v corresponds to $gp_{(j \oplus i)}^{[v']}$ of v' ($i = 1$ in the example). Hence, we compare the angle between the generating points $gp_j^{[v]}$ and $gp_{(j \oplus 1)}^{[v]}$, $\alpha_j^{[v]}$, with the corresponding angle $\alpha_{(j \oplus i)}^{[v']}$. For the corresponding angles the difference needs to be under a given error tolerance γ_a in order to be considered compatible. The same goes for the difference in the radii of inscribed circles resulting in the following compatibility criterion:

$$\text{compatible}_{\text{localgeom}}(v, v') \Leftrightarrow \left(\forall j, 1 \leq j \leq \deg(v) : |\alpha_j^{[v]} - \alpha_{(j \oplus i)}^{[v']}| < \gamma_a \right) \tag{14}$$
$$\wedge\, |r(v) - r(v')| < \gamma_{\text{radius}}$$

---

[2] For simplicity, we equate nodes with their respective feature vectors here.

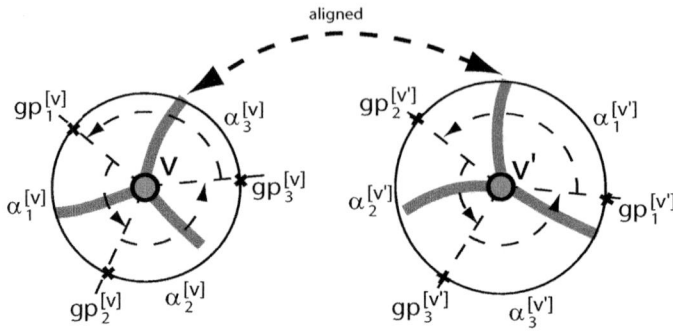

Fig. 8. Comparison of the local geometry of Voronoi nodes: The given edge alignment allows the comparison of the angles between corresponding generating points. In addition, the difference between the radii of the maximal inscribed circles is taken into account.

We subsume the unary constraints defined in Eq. 12–14 in this section under a single compatibility criterion for nodes in a subtree called $\text{compatible}_{\text{unary}}(v, v')$. It is incorporated into the overall edit distance computation by including it into the definitions of $\text{dist}_{\text{aligned}}(u, u', i)$ (Eq. 8) and $\text{dist}_m(\triangle_v^{(u)}, \triangle_{v'}^{(u')})$ (Eq. 4) so that the respective function yields ∞ when the criterion is violated. All other equations remain unchanged.

5.2   Binary and Ternary Constraints Based on Relative Position

We describe the geometric configuration of nodes in an AGVG in terms of distances between adjacent nodes and the angles formed at the nodes by the straight edges. While this approach has the downside that it is less strict than comparing relative distances for each pair of matched features (see for instance [33]) because error tolerances can add up along the path connecting two nodes, it has one crucial advantage: The first approach cannot not be employed without breaking the optimal substructure condition underlying the dynamic programming approach. The best matching of a subtree here does not necessarily lead to an optimal solution to the overall problem because the best matching for the overall solution now depends on the compatibility with matches made at a higher level of the compared trees.

In our approach, on the other hand, the dependency is restricted to the next higher matching (in the case of binary distance constraints) or the next two higher matchings (in the case of ternary angle constraints) as we will see in the following. This allows an extension of our approach that increases the worst-case complexity to a much lesser degree.

Our binary criterion restricting pairs of matchings $u \rightsquigarrow u'$ and $v \rightsquigarrow v'$ describes the distance between neighboring nodes. This happens both in terms of their line-of-sight distance l_{los} derived from the edge lengths (and possible inter-mediate angles) and the distance l_{course} along the Voronoi curves derived from the course information and stored as attributes of the edges.

The binary distance criterion which we simply call compatible$_{binary}$ is then given by:

$$\text{compatible}_{binary}((u, u'), (v, v')) \Leftrightarrow |l_{los}(u, v) - l_{los}(u', v')| < \gamma_{l_{los}} \qquad (15)$$
$$\wedge |l_{course}(u, v) - l_{course}(u', v')| < \gamma_{l_{course}}$$

In the following, we want to include the angle $\beta(u, v, w)$ formed by connecting a node v with one of its ancestors u and one of its child nodes w into the matching scheme. Hence, we are now dealing with ternary constraints involving three pairs of matched nodes $u \rightsquigarrow u'$, $v \rightsquigarrow v'$ and $w \rightsquigarrow w'$.

We define the criterion for accepting such a matching in a way that it is still true if either $u = v$ or $u' = v'$ and, hence, no angle is defined for one of the AGVGs. However, we explicitly demand that $v \neq w$ and $v' \neq w'$. The reasons for this will become clear in a moment. The ternary compatibility then is:

$$\text{compatible}_{ternary}((u, u'), (v, v'), (w, w')) \Leftrightarrow u = v \vee u' = v' \vee$$
$$|\beta(u, v, w) - \beta(u', v', w')| < \gamma_{angle}$$

$$(16)$$

As we mentioned previously, even for the chosen approach of incorporating con-figurational information, we have to modify our matching approach because the best matching of two subtrees for the overall solution now depends on the previ-ous node matching (for the binary distance constraint) and on the previous two node matchings for the ternary angle constraints. Consequentially, we turn from simple subtrees to matching more complex subtree structures which contain two node pairs (u, u') and (v, v') considered as matched. We call these structures *mm-subtrees* and they consist of:

1. a root node u (respectively u') considered as matched,
2. a first sequence of edges whose intermediate nodes are considered as deleted (or added),
3. another node v (respectively v') considered as matched,
4. a second sequence of edges with intermediate deleted (or added) nodes,
5. a subtree with the currently considered node w (respectively w') as its root.

An example of a mm-subtree matching is shown in Fig. 9. We denote mm-subtrees as $\triangle_w^{u,v}$. Admitting that the matched nodes u and v are identically allows us to handle cases in which only one matched node pair exists so far (when starting the matching for the chosen start nodes of the AGVGs in dist$_{node}$ and dist$_{aligned}$) and in which, hence, only the binary direction constraint can be applied.

Modifying the edit distance equations by replacing subtrees by mm-subtrees and including the compatible$_{ternary}$ criterion yields our final matching algorithm.

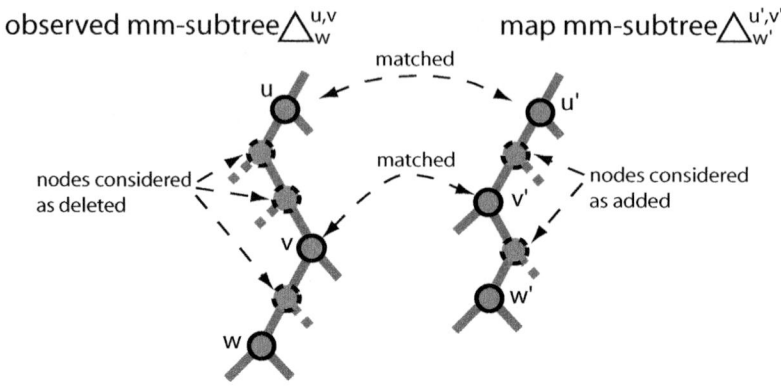

Fig. 9. Matching two mm-subtrees

Below is the modified definition (indicated by the three primes) of the edit distance equations for the match operation:

$$\text{dist}'''_{\text{match}}(\triangle_w^{u,v}, \triangle_{w'}^{u',v'}) =$$
$$\begin{cases} \sum_{i=1}^{deg(w)-1} \text{dist}'''_{\text{st}}\left(\triangle_{\text{child}_i(w)}^{v,w}, \triangle_{\text{child}_i(w')}^{v,w}\right), & \text{compatible}(u,v) \\ & \wedge \text{compatible}((v,v'),(w,w')) \\ & \wedge \text{compatible}_{\text{ternary}}((u,u'),(v,v'),(w,w')) \\ \infty & , \text{else} \end{cases}$$
(17)

For the other two operations we simply need to replace subtrees by mm-subtrees to get the definitions of $\text{dist}'''_{\text{d}}$ and $\text{dist}'''_{\text{a}}$. The same goes for $\text{dist}'''_{\text{st}}$ replacing dist_{st} and $\text{dist}'''_{\text{node}}$ replacing $\text{dist}_{\text{node}}$.

$\text{dist}'''_{\text{aligned}}$ is the mentioned special case where the computation starts with mm-subtrees with identical u and v. No parent node exists and, hence, no angle will be checked here. Including an angle constraint for the root node would make the optimal solutions for subtrees of u dependent on the matchings in its other subtrees breaking the optimal substructure condition.

$$\text{dist}'''_{\text{aligned}}(u, u', i) =$$
$$\begin{cases} \sum_{j=1}^{deg(u)} \text{dist}'_{\text{subtree}}\left(\triangle_{\text{neighbor}_j(u)}^{u,u}, \triangle_{\text{neighbor}_{(j\oplus i)}(u')}^{u',u'}\right), & \text{compatible}_{\text{unary}}(u,u') \\ \infty & , \text{else} \end{cases}$$
(18)

Finally, the overall edit distance $\text{dist}'''_{\text{AGVG}}$ remains unchanged except for making use of $\text{dist}'''_{\text{node}}$ instead of $\text{dist}_{\text{node}}$.

With regard to complexity, the change from subtrees to mm-subtrees means that the size of a table containing all $\text{dist}'''_{\text{st}}$ values now depends cubicly on m

and n. The number of mm-subtrees $\triangle_w^{u;v}$ with $u = v$ in a graph with n nodes is $n(n+1)$. The number of mm-subtrees with $u \neq v \neq w$ depends on the structure of the graph. In the worst case, the graph is a linear chain. In this case, we get $\frac{n(n+1)(2n+1)}{6} - \frac{n(n+1)}{2}$ additional mm-subtrees. As a result, the overall number of mm-subtrees in the worst-case is of order $O(n^3)$ and we get an upper boundary on the time complexity of $O(m^3 n^3)$. However, AGVGs typically are not linear chains but approximately 3-regular graphs.

Possible applications of the data association algorithm for AGVGs developed in this section are the tracking of Voronoi nodes, localization within a given map AGVG, or incrementally adding new information to the map (as will be discussed in the next section). However, on its own it does not offer a sufficient solution to the problem of constructing a global graph representation as it will not close loops in the environment. To achieve this, the matching would have to be extended or the approach needs to be embedded into a global robot mapping approach. The experimental analysis in the Sect. 7 will show that the matching is much more efficient in practice than could be expected from the theoretically determined worst-case time complexity of $O(m^3 n^3)$ which makes this embedding feasible.

6   Map Merging Based on the Matching Result

Once the cost-optimal matching has been computed using the approach described in the previous two sections, it can be used to incorporate newly perceived subgraphs into the map. This can be done straightforwardly by following the operations that led to the minimal edit distance. When computing the edit distance, we therefore not only store the computed $\text{dist}_{\text{st}}'''$ values but also the operation chosen for this particular subproblem. When this operation is a delete or add operation, we also record the i for which the minimal distance has been achieved (see Eqs. 5 and 6). In addition to this, we need to keep the pair of start nodes and the edge offset used in the best matching to recover the complete sequence of operations (see Eqs. 7 and 9).

Incorporating new subgraphs into the model AGVG has to take place for every delete operation as this means that subtrees contained in the local AGVG are not contained in the map. The resulting map transformation algorithm works analogously to the edit distance computation. We provide a pseudocode version in Algorithm 2. The procedure names have been chosen in accordance with the names of the edit distance functions. The main procedure transformAGVG basically combines $\text{dist}_{\text{AGVG}}'''$, $\text{dist}_{\text{node}}'''$, and $\text{dist}_{\text{aligned}}'''$. In addition to new parts being added to the model AGVG, information about already known features which have been matched can be updated (e.g., updating the position estimates of the nodes). This happens in line 3 of transformAGVG and in line 1 of the transformMatch function.

In Fig. 10, we see an example of applying the map transformation procedure. The dark AGVG in Fig. 10(a) is the map AGVG and the other AGVG is the local AGVG. The optimal matching computed by our algorithm is given by the two matched node pairs connected by the small ellipses, and the edit operations

---

Algorithm 2. Transformation algorithm for local AGVG G and map AGVG G'

---

procedure transformAGVG(AGVG G, AGVG G')

1: $v, v' \leftarrow v \in G, v' \in G'$ for which $\text{dist}'''_{\text{node}}(v, v')$ was minimal
2: $o \leftarrow$ edge offset which resulted in minimal costs
3: update v'
4: **for** $i = 1$ to $\deg(v)$ **do**
5:     transformSubtree($\triangle^{v,v}_{\text{child}_i(v)}, \triangle^{v',v'}_{\text{child}_{i \oplus o}(v')}$)
6: **end for**

procedure transformSubtree($\triangle^{u,v}_w, \triangle^{u',v'}_{w'}$)

1: **if** operation($\triangle^{u,v}_w, \triangle^{u',v'}_{w'}$) = *match* **then**
2:     transformMatch($\triangle^{u,v}_w, \triangle^{u',v'}_{w'}$)
3: **else if** operation($\triangle^{u,v}_w, \triangle^{u',v'}_{w'}$) = *delete* **then**
4:     transformDelete($\triangle^{u,v}_w, \triangle^{u',v'}_{w'}$)
5: **else if** operation($\triangle^{u,v}_w, \triangle^{u',v'}_{w'}$) = *add* **then**
6:     transformAdd($\triangle^{u,v}_w, \triangle^{u',v'}_{w'}$)
7: **end if**

procedure transformMatch($\triangle^{u,v}_w, \triangle^{u',v'}_{w'}$)

1: update w'
2: **for** $i = 1$ to $\deg(w)$ **do**
3:     transformSubtree($\triangle^{v,w}_{\text{child}_i(w)}, \triangle^{v',w'}_{\text{child}_i(w')}$)
4: **end for**

procedure transformDelete($\triangle^{u,v}_w, \triangle^{u',v'}_{w'}$)

1: $p \leftarrow i$ which resulted in minimal costs for $\triangle^{u,v}_w, \triangle^{u',v'}_{w'}$
2: add w' with child$_j$ subtrees, $j \neq p$, between v' and w' into $\triangle^{u',v'}_{w'}$
3: transformSubtree($\triangle^{u,v}_{\text{child}_p(w)}, \triangle^{u',v'}_{w'}$)

procedure transformAdd($\triangle^{u,v}_w, \triangle^{u',v'}_{w'}$)

1: $p \leftarrow i$ which resulted in minimal costs for $\triangle^{u,v}_w, \triangle^{u',v'}_{w'}$
2: transformSubtree($\triangle^{u,v}_w, \triangle^{u',v'}_{\text{child}_p(w')}$)

---

are marked by the letters 'a' for add and 'd' for delete. In Fig. 10(b), we see how the model AGVG has been complemented with the additional branches from the local AGVG. The basic update approach sketched here can be further extended by incorporating improved techniques for deciding which features should be kept in the map. For instance, an add operation contained in the optimal matching is an indication that a node contained in the map may have been a spurious one that should better be removed.

7   Experiments

In this section, we summarize experiments we performed to evaluate the efficiency of the relevance value computation described in Sect. 3.3 and in particular the performance of our AGVG matching approach developed in Sects. 4 and 5.

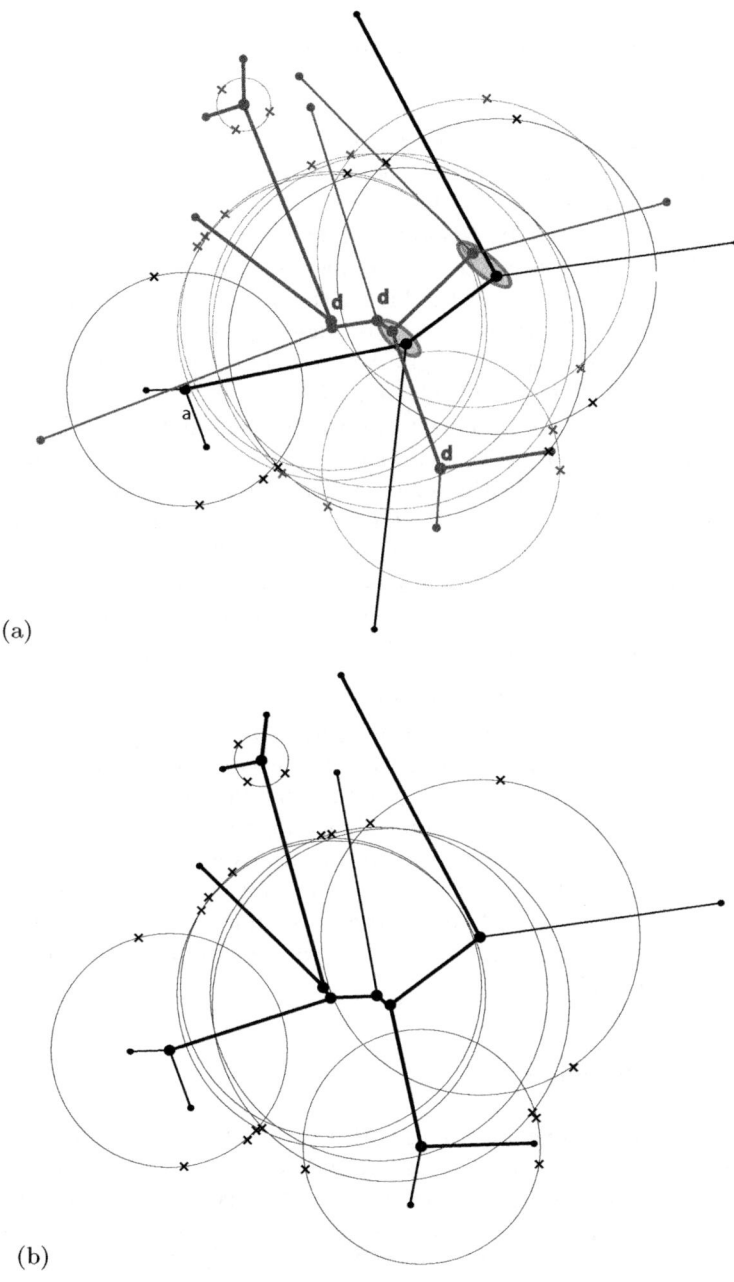

(a)

(b)

Fig. 10. Map transformation based on the optimal matching: (a) The small ellipses show the two matched node pairs, and the letters 'd' and 'a' mark where the corresponding edit operation has been performed. (b) shows the complemented map AGVG.

7.1   Relevance Value Computation

To empirically evaluate the efficiency of the relevance value computation, we implemented a random graph generator that produces pseudo-AGVGs. It first produces a 3-regular graph for a given number of edges. The nodes are placed on a hexagonal grid and each node is connected with its three neighbors in the grid. In a second step the number of nodes is varied by splitting nodes of degree 3 into three nodes of degree 1 while ensuring that the graph remains connected. This procedure allows us to generate AGVGs with a desired edge-node ratio. This ratio is directly related to the number of cycles appearing in the graph: the higher the edge-node ratio, the higher the number of cycles in the graph. In a final step, the length attributes of the edges and the radius attributes of the nodes are varied randomly. The edge-node ratio of the generated pseudo-AGVGs lies between 1 (tree-formed AGVG) and 1.5 (3-regular AGVG).

In our comparison, we systematically varied the number of edges and nodes in the randomly generated AGVGs and applied both the basic variant of the relevance computation algorithm given in Algorithm 1 and the improved variant including the optimizations described in Sect. 3.4 to the resulting graphs. The resulting computation times on a 2 GHz Pentium M CPU are shown in Figs. 11(a) and 11(c).[3] All data points are averages taken over 30 different pseudo-AGVGs. Figs 11(b) and 11(d) each show two cuts through the data sets for fixed edge-node ratios, one for a high ratio ($\frac{|E|}{|V|} > 1.47$) and one for a ratio close to 1 ($\frac{|E|}{|V|} < 1.1$). As the data shows, the computation time of the basic version of the value computation algorithm increases significantly when the ratio of edges to nodes goes towards 1. The reason is that in this case the graphs contain almost no cycles and almost the complete graph is expanded for each node.

In contrast, the improved version only shows a slight increase towards lower edge-node ratios because, due to the optimizations made, typically only a small part of the graph needs to be considered even for tree-like structures. Overall, the computation times stayed below 40 ms even for very large graphs with 1 000 nodes and edges. As a result, the improved algorithm is efficient enough for typical mapping applications.

7.2   AGVG Matching

We evaluated the performance of our AGVG matching approach by comparing its performance with that of the standard ICNN data association approach which we modified by incorporating our unary compatibility criterion compatible$_{unary}$. The goal of this evaluation was to determine the differences in the quality of the computed associations and the required computation times. As one advantage of the AGVG matching approach is that, as a batch association method, it is

---

[3] Both diagrams show an increasing number of gaps for higher edge and node numbers and an edge-node ratio close to 1. These are cases in which the random graph generator failed to create suitable instances in reasonable time.

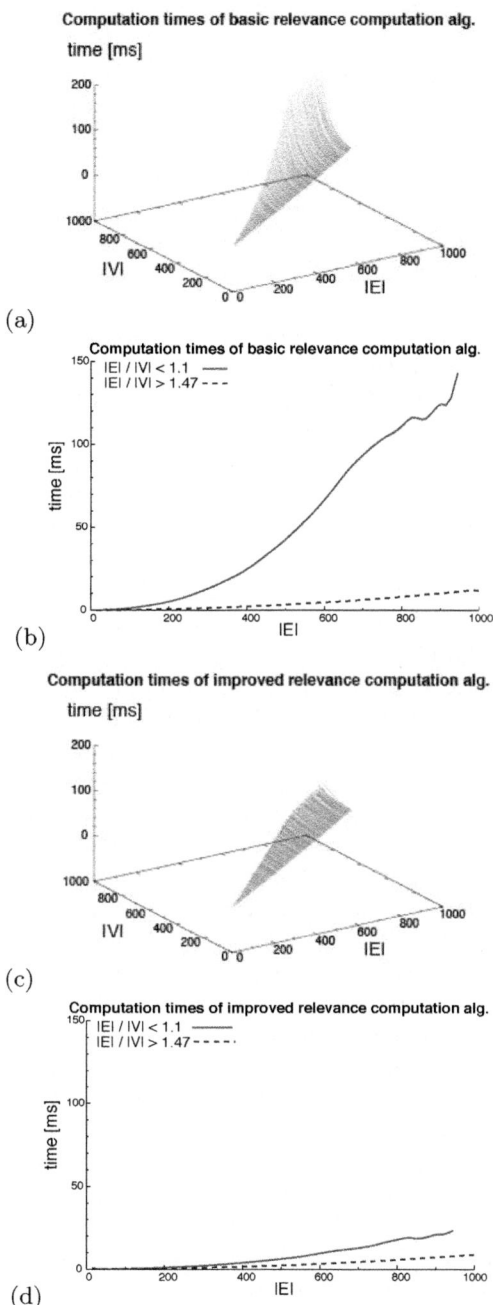

Fig. 11. Computation times for (a) the basic relevance value computation algorithm and (c) the optimized version depending on the number of nodes and edges in the randomly generated pseudo-AGVGs. (b) and (d) show the curves for AGVGs with a high edge-node ratio and graphs with an edge-node ratio close to 1.

not completely reliant on pose estimates, we also evaluate its performance when no pose information is available.

For the comparison of ICNN and AGVG matching, we incorporated both data association algorithms into a feature-based FastSLAM algorithm (see [38]) based on Voronoi nodes, confronting both algorithms with a large amount of data association problems like the one shown in Fig. 6. Ground truth information about the identity of the Voronoi nodes was manually added to the local AGVGs. Every time a data association needs to be computed, this is done for both, ICNN and our AGVG matching, and the results are stored together with the required computation times.

In our experiments, we ran the algorithm for two different environments using different start positions in the data sets. The local maps contained on average 19 Voronoi nodes. The size of the map AGVG varied between 9 and 588 nodes.

To compare the data association quality of the two algorithms, we recorded the number of wrong assignments in each data association. As correct associations are more critical when they concern more relevant Voronoi nodes, we also determined the average vnrm values of the wrongly associated Voronoi nodes. In addition, we recorded the computation time for each data association. For the AGVG matching algorithm we also determined the fraction of entries in the edit distance table that actually had to be calculated. The experiment was performed on a 2GHz Pentium M CPU.

The result of this experiment was that AGVG matching shows a significantly increased data association quality. 94.67% of all decisions are correct in contrast to 71.50% for the modified ICNN algorithm. Additionally, we determined the average relevance value of wrongly associated nodes which is 0.19 for our approach compared to 0.55 for ICNN. This means the errors are made for more irrelevant nodes. In particular, we observed that AGVG matching never made wrong associations for the most relevant Voronoi nodes of the local map, while ICNN frequently inverts the assignments of very relevant Voronoi nodes positioned close to each other.

While an improved data association quality was to be expected because AGVG matching is a batch association technique whereas ICNN makes individual assignments, the more important question is at which computational costs this improvement can been achieved. Fig. 12 depicts the average computation times over the size of the map AGVG (given by the number of Voronoi nodes). It turns out that the increase in computational costs is surprisingly low. In contrast to the theoretical upper bound of cubic dependency on the size of the map AGVG, the increase in computation time was almost linear in the investigated range of map sizes. This shows the effectiveness of the constraint-based pruning and the employed branch and bound techniques. On average, only 0.00001% of the entries in the edit distance table had to be computed.

We then repeated the previous experiment for the AGVG matching algorithm, but this time without employing the pose information in the matching algorithm. This means that the compatibility criteria $compatible_{absdist}$ and $compatible_{absdist\text{-}gp}$ were removed from $compatible_{unary}$. The result of this experiment was that in 87.8% of all cases the AGVG matching approach deter-mined the same data association as with pose information and, hence, was able to localize the robot correctly. In absence of pose information, AGVG matching still achieved 83.21% correct assignments. As a result, the approach is also well-suited for global localization or map merging applications. We observed that cases in which the matching resulted in an entirely wrong localization typically involved local AGVGs with very few nodes leading to a high level of ambiguity. The average computation times shown in Fig. 13 were noticeably higher than in the first experiment as more possibilities needed to be explored. However, the approach still scaled well with the size of the map AGVG. The average percent-age of calculated entries in the edit distance table increased to 0.00009% but remained very low in general.

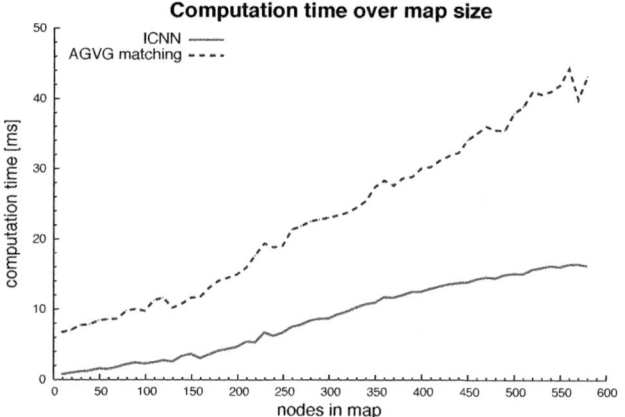

Fig. 12. Comparison of average computation times of ICNN and AGVG matching depending on the number of nodes in the map AGVG

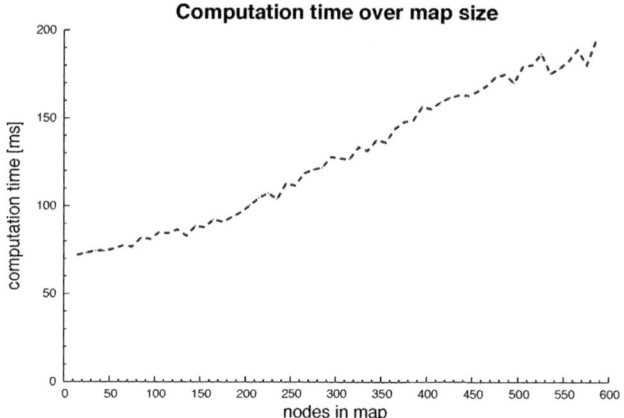

Fig. 13. Average computation times of AGVG matching without pose information

Overall, the experiments we described in this section demonstrate that the AGVG matching approach achieves an excellent ratio between quality and computational cost. As a result, the approach is not only well-suited when a high level of reliability in the data association is required (e.g., in single-hypothesis Kalman filter based approaches [39,40]) but also when computational efficiency is required as in a particle filter-based mapping approach [41,38] in which multiple data associations need to be determined in each update step. Furthermore, the exploitation of configurational information makes it applicable for tasks in which no pose information is available.

8   Conclusions

We developed an AGVG matching algorithm to tackle the data association problem in a robot navigation and mapping scenario in which generalized Voronoi graphs are used as the robot's spatial representation of the environment. Our approach employs inexact tree matching techniques based on edit distance and heavily exploits additional spatial constraints based on node and edge attributes. We based the edit operations on a model of relevance and stability of the nodes and edges in an AGVG and showed that these relevance values can be computed efficiently. The matching algorithm has been realized as a top-down dynamic programming approach making extensive use of branch and bound techniques. The evaluation of our algorithm showed that it yields very low error rates with only a modest increase in computational costs compared to the standard ICNN approach. Hence, the algorithm is well-suited for localization and map building. However, the underlying tree matching approach itself currently does not facilitate loop closing and therefore needs to be embedded into a global mapping system. Extending the algorithm to also cover loop closing will be a topic for future research.

Acknowledgments. The author would like to thank Christian Freksa and Lutz Frommberger for valuable discussions. Several anonymous reviewers provided helpful comments on earlier versions of the presented material [42]. Funding by the Deutsche Forschungsgemeinschaft (DFG) under grants SFB/TR 8 Spatial Cognition and IRTG GRK 1498 Semantic Integration of Geospatial Information is gratefully acknowledged.

References

1. Aurenhammer, F.: Voronoi diagrams – A survey of a fundamental geometric data structure. ACM Computing Surveys 23(3), 345–405 (1991)
2. Okabe, A., Sugihara, K., Chiu, S.N., Boots, B.: Spatial Tessellations - Concepts and Applications of Voronoi Diagrams. John Wiley and Sons, Chichester (2000)
3. Lee, D.T., Drysdale III, R.L.S.: Generalization of Voronoi diagrams in the plane. SIAM Journal on Computing 10(1), 73–87 (1981)

4. Kirkpatrick, D.G.: Efficient computation of continuous skeletons. In: Annual IEEE Symposium on Foundations of Computer Science, pp. 18–27 (1979)

5. Ó'Dúnlaing, C., Yap, C.K.: A retraction method for planning the motion of a disc. Journal of Algorithms 6, 104–111 (1982)

6. Latombe, J.C.: Robot Motion Planning. Kluwer Academic Publishers, Dordrecht (1991)

7. Remolina, E., Kuipers, B.: Towards a general theory of topological maps. Artificial Intelligence 152(1), 47–104 (2004)

8. Kuipers, B.: The Spatial Semantic Hierarchy. Artificial Intelligence 119(1-2), 191–233 (2000)

9. Choset, H., Burdick, J.: Sensor-based exploration: The Hierarchical Generalized Voronoi Graph. The International Journal of Robotics Research 19(2), 96–125 (2000)

10. Beeson, P., Jong, N.K., Kuipers, B.: Towards autonomous topological place detection using the Extended Voronoi Graph. In: IEEE International Conference on Robotics and Automation (ICRA 2005), pp. 4373–4379 (2005)

11. Moravec, H., Elfes, A.: High resolution maps from angle sonar. In: Proceedings of the IEEE Conference on Robotics and Automation (ICRA 1985), pp. 116–121 (1985)

12. Crowley, J.: World modeling and position estimation for a mobile robot using ultrasonic ranging. In: Proceedings of IEEE International Conference on Robotics and Automation (ICRA 1989), pp. 674–680 (1989)

13. Leonard, J.J., Durrant-Whyte, H.F.: Simultaneous map building and localization for an autonomous mobile robot. In: Proceedings of IEEE/RSJ International Workshop on Intelligent Robots and Systems, pp. 1442–1447 (1991)

14. Thrun, S., Burgard, W., Fox, D.: Probabilistic Robotics. MIT Press, Cambridge (2005)

15. Bar-Shalom, Y., Fortmann, T.E.: Tracking and Data Association. Academic Press, London (1988)

16. Grimson, W.E.L.: Object Recognition by Computer – The Role of Geometric Constraints. MIT Press, Cambridge (1990)

17. Bailey, T., Nieto, J., Nebot, E.: Consistency of the FastSLAM algorithm. In: IEEE International Conference on Robotics and Automation (ICRA 2006), pp. 424–429 (2006)

18. Wallgrün, J.O.: Autonomous construction of hierarchical Voronoi-based route graph representations. In: Freksa, C., Knauff, M., Krieg-Brückner, B., Nebel, B., Barkowsky, T. (eds.) Spatial Cognition IV. LNCS (LNAI), vol. 3343, pp. 413–433. Springer, Heidelberg (2005)

19. Sanfeliu, A., Fu, K.: A distance measure between attributed relational graph. IEEE Transactions on Systems, Man and Cybernetics 13, 353–362 (1983)

20. Eshera, M.A., Fu, K.S.: An image understanding system using attributed symbolic representation and inexact graph-matching. IEEE Transactions on Pattern Analysis and Machine Intelligence 8(5), 604–618 (1986)

21. Zhang, K., Shasha, D.: Simple fast algorithms for the editing distance between trees and related problems. SIAM Journal on Computing 18(6), 1245–1262 (1989)

22. Zhang, K., Statman, R., Shasha, D.: On the editing distance between unordered labeled trees. Information Processing Letters 42(3), 133–139 (1992)

23. Sebastian, T.B., Klein, P.N., Kimia, B.B.: Recognition of shapes by editing their shock graphs. IEEE Transactions on Pattern Analysis and Machine Intelligence 26(5), 550–571 (2004)

24. Blum, H.: A transformation for extracting new descriptors of shape. In: Wathen-Dunn, W. (ed.) Models for the Perception of Speech and Visual Form, pp. 362–381. MIT Press, Cambridge (1967)
25. Krieg-Brückner, B., Frese, U., Lüttich, K., Mandel, C., Mossakowski, T., Ross, R.: Specification of an Ontology for Route Graphs. In: Freksa, C., Knauff, M., Krieg-Brückner, B., Nebel, B., Barkowsky, T. (eds.) Spatial Cognition IV. LNCS (LNAI), vol. 3343, pp. 390–412. Springer, Heidelberg (2005)
26. Wallgrün, J.O.: Hierarchical Voronoi Graphs – Spatial Representation and Reasoning for Mobile Robots. Springer, Heidelberg (2009)
27. Ogniewicz, R.L., Kübler, O.: Hierarchic Voronoi Skeletons. Pattern Recognition 28(3), 343–359 (1995)
28. Siddiqi, K., Kimia, B.B.: A shock grammar for recognition. In: IEEE Conference on Computer Vision and Pattern Recognition, pp. 507–513 (1996)
29. Mayya, N., Rajan, V.T.: Voronoi diagrams of polygons: A framework for shape representation. Journal of Mathematical Imaging and Vision 6(4), 355–378 (1996)
30. Dijkstra, E.W.: A note on two problems in connexion with graphs. Numerische Mathematik 1, 269–271 (1959)
31. Neira, J., Tardós, J.D.: Data association in stochastic mapping using the joint compability test. IEEE Transactions on Robotics and Automation 17, 890–897 (2001)
32. Mahalanobis, P.: On the generalized distance in statistics. Proceedings of the National Institute of Sciences of India 12, 49–55 (1936)
33. Bailey, T.: Mobile Robot Localisation and Mapping in Extensive Outdoor Environments. PhD thesis, University of Sydney (2001)
34. Wolter, D.: Spatial Representation and Reasoning for Robot Mapping - A Shape-Based Approach. Springer Tracts in Advanced Robotics, vol. 48. Springer, Heidelberg (2008)
35. Lim, J.H., Leonard, J.J.: Mobile robot relocation from echolocation constraints. IEEE Transactions on Pattern Analysis and Machine Intelligence 22(9), 1035–1041 (2000)
36. Arras, K., Castellanos, J., Schilt, M., Siegwart, R.: Feature-based multi-hypothesis localization and tracking using geometric constraints. Robotics and Autonomous Systems Journal 44(1) (2003)
37. Aho, A., Hopcroft, J., Ullman, J.: The Design and Analysis of Computer Algorithms. Addison-Wesley, Reading (1974)
38. Montemerlo, M., Thrun, S., Koller, D., Wegbreit, B.: FastSLAM: A factored solution to the simultaneous localization and mapping problem. In: Proceedings of the AAAI National Conference on Artificial Intelligence, pp. 593–598 (2002)
39. Smith, R.C., Cheeseman, P.: On the representation and estimation of spatial uncertainty. The International Journal of Robotics Research 5(4), 56–68 (1986)
40. Dissanayake, M.G., Newman, P., Clark, S., Durrant-Whyte, H., Csorba, M.: A solution to the simultaneous localization and map building (SLAM) problem. IEEE Transactions on Robotics and Automation 17(3), 229–241 (2001)
41. Murphy, K.: Bayesian map learning in dynamic environments. In: Solla, S.A., Leen, T.K., Müller, K.R. (eds.) Advances in Neural Information Processing Systems 12, pp. 1015–1021. The MIT Press, Cambridge (2000)
42. Wallgrün, J.O.: Matching annotated generalized Voronoi graphs for autonomous robot localization and mapping. In: Anton, F., Bærentzen, J.A. (eds.) Proceedings of the 6th Annual International Symposium on Voronoi Diagrams in Science and Engineering. IEEE Computer Society, Los Alamitos (2009)

Properties and an Approximation Algorithm of Round-Tour Voronoi Diagrams

Hidenori Fujii[1] and Kokichi Sugihara[2]

[1] Department of Mathematical Informatics, University of Tokyo,
Tokyo 113-8656, Japan
fujii@ipl.t.u-tokyo.ac.jp
[2] Meiji Institute for Advanced Study of Mathematical Sciences, Meiji University,
Kawasaki 214-8571, Japan
kokichis@isc.meiji.ac.jp

Abstract. This paper proposes a new generalization of the Voronoi diagram. Consider two kinds of facilities located in a city, for example, restaurants and bookstores. We want to visit both and return to our house. To each pair of a restaurant and a bookstore is assigned a region such that a resident in this region can visit them in a shorter round tour than visiting any other pair. The city is partitioned into these regions according to which pair of a restaurant and bookstore permits the shortest round tour. We call this partitioning a "round-tour Voronoi Diagram" for the restaurants and bookstores. We study the basic properties of this Voronoi diagram and consider an efficient algorithm for its approximate construction.

Keywords: Generalized Voronoi diagram; round-tour; restaurants and bookstores; facility location analysis; shortest round tour.

1 Introduction

The Voronoi diagram is one of the most fundamental geometric data structures because of its fruitful generalizations and applications [15]. Generalization of the Voronoi diagram can be taken in three directions.

The first direction comprises generalizations of the distance. The Euclidean distance, which is used for the ordinary Voronoi diagram, can be replaced by the L_p distance [6], the collision-avoidance distance [1], the power distance [4,10], the weighted distance [3], or the boat-sail distance [12], to mention a few.

The second direction comprises generalizations of the underlying space. The ordinary Voronoi diagram is in Euclidean space; however, the space can be replaced by a spherical surface [18], polygonal surface [2], or network [8].

The third direction comprises generalizations of the generators. The ordinary Voronoi diagram is defined for points; however, they can be replaced by general figures such as circles, line segments, and polygons [17], or replaced by subsets of points in higher-order Voronoi diagrams [16].

In this paper, we propose a new generalization of the Voronoi diagram [7]. Our generalization might be understood easily in the context of a round tour involving

M.L. Gavrilova et al. (Eds.): Trans. on Comput. Sci. IX, LNCS 6290, pp. 109–122, 2010.

a visit to both a restaurant and bookstore before returning home. Suppose that there are many restaurants and bookstores in a city. The city is partitioned into regions according to the pair of restaurant and bookstore one can visit in the shortest round tour. We call this partition the "round-tour Voronoi diagram" for the restaurants and bookstores. We study the basic properties of this diagram and consider an efficient algorithm for computing the approximation of this diagram in the form of a digital picture.

This work is closely related to Ohyama's work [13,14] in which consumer behavior is studied using his new Voronoi diagrams. He considered a consumer who visits stores one by one until he selects a good. In that case, the "distance" from a point (where the consumer lives) to a set of stores is the expected length of the shortest path along which he travels while shopping.

Gill Barequet et al. [5] study a different but related topic. In their paper, they define a new type of a planar distance functions from a point to a pair of points. They focus on a few such distance functions, analyze the structure and complexity of the corresponding nearest- and furthest-neighbor Voronoi diagrams (in which every region is defined by a pair of point sites), and show how to compute their diagrams efficiently. Our new Voronoi diagram is different from theirs in that the generating points are divided into two groups and a pair of points are generated by taking one from each group. Very recently, Hanniel and Barequet [9] concentrated on one special version of their Voronoi diagrams called the triangle-perimeter two-site Voronoi diagram, which is very similar to ours, with only one difference that there is only one type of generators. This difference results in big difference in the number of Voronoi regions. We will discuss this in setion 3.

In section 2 we introduce our new Voronoi diagram, the round-tour Voronoi diagram, and in section 3 we consider its basic properties. In sections 4 and 5, we construct an algorithm for computing a digital approximation of the Voronoi diagram, and in section 6 we give concluding remarks.

2   Round-Tour Voronoi Diagram for Two Sets of Generators

Let $A = \{a_1, a_2, \ldots, a_n\}$ be a set of n points in the plane $\mathbb{R}^2$. For any two points $x, y \in \mathbb{R}^2$, let $d(x, y)$ denote the Euclidean distance between x and y. We define $V(A; a_i)$ as

$$V(A; a_i) = \{z \in \mathbb{R}^2 \mid d(z, a_i) < d(z, a_j), j \neq i\}. \tag{1}$$

$V(A; a_i)$ represents the set of points that are closer to a_i than to any other point in A. The plane is partitioned into $V(A; a_1), V(A; a_2), \ldots, V(A; a_n)$ and their boundaries. We call this partition *the Voronoi diagram* for A, and $V(A; a_i)$ *the Voronoi region* of a_i. The elements of A are called *generators* (or *generating points*) of the Voronoi diagram. This is the definition of the ordinary Voronoi diagram.

Suppose that there are many restaurants and bookstores in a city. Let A be the set of points at which the restaurants are located, and B be that at which the bookstores are located. We assume that all restaurants are identical in the sense that people do not have any preference except for their distances, and that all the bookstores are identical in a similar sense. Figure 1 shows an example of this situation, where small empty circles represent the restaurants and the small filled squares represent the bookstores. The solid lines in this figure show the Voronoi diagram for A, and the broken lines show the Voronoi diagram for B.

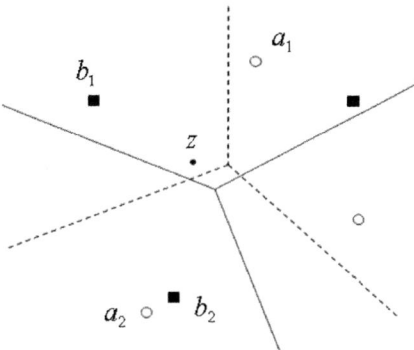

Fig. 1. Overlay of two Voronoi diagrams: the solid lines for A and the broken lines for B

Consider a person who lives at z. If he/she wants to go to a restaurant, it is best to go to a_1 because z belongs to the Voronoi region of a_1 in the Voronoi diagram for A (i.e., the solid-line diagram), that is, a_1 is the nearest restaurant from z.

Similarly, if he/she wants to go to a bookstore, it is most convenient to go to bookstore b_1 because z belongs to the Voronoi region of b_1 in the Voronoi diagram for B (i.e., the broken-line diagram).

Next, suppose that a person at z wants to visit both a restaurant and a bookstore in one tour. Then, a_1 and b_1 are not a good pair, because the person can visit restaurant a_2 and bookstore b_2 in a shorter round tour. Thus, in this context, the ordinary Voronoi diagrams do not give us enough information for a resident to choose a restaurant and a bookstore. This observation natually reads us to the next generalization of the Voronoi diagram.

Let z be a general point on the plane. For $a \in A$ and $b \in B$, we define

$$l_z(a, b) = d(z, a) + d(a, b) + d(b, z). \tag{2}$$

The value $l_z(a, b)$ is the length of the perimeter of the triangle formed by three vertices a, b, and z; that is, $l_z(a, b)$ is the length of the shortest round tour starting at z, visiting a and b, and returning to z.

For $a \in A$ and $b \in B$, let us define $V(A, B; a, b)$ by

$$V(A, B; a, b) = \{z \in \mathbb{R}^2 \mid l_z(a, b) = \min_{a' \in A, \, b' \in B} l_z(a', b')\}. \tag{3}$$

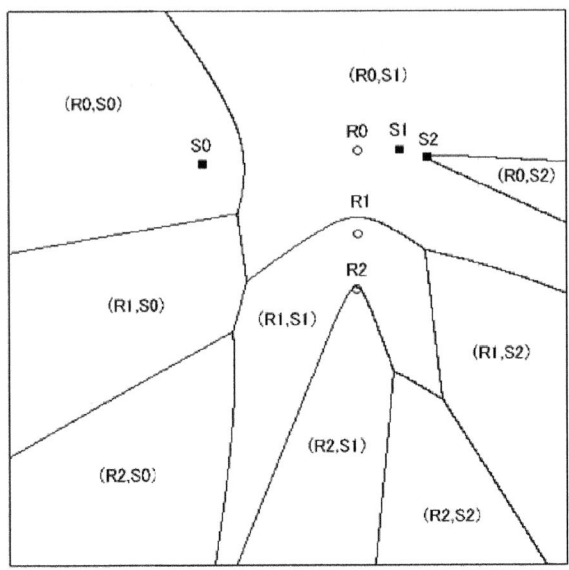

Fig. 2. Example of a round-tour Voronoi diagram

Intuitively, $V(A, B; a, b)$ represents the region in which any resident can visit a and b in a shorter round tour than he/she visits other pairs of a restaurant and bookstore.

The plane is decomposed into the regions $V(A, B; a, b)$ (where $a \in A$ and $b \in B$) without overlap except for the boundaries. We call this partition the *round-tour Voronoi diagram* for A and B. A and B are called the *generating sets* of the Voronoi diagram.

Figure 2 shows an example of the round-tour Voronoi diagram for three restaurants $A = \{R_0, R_1, R_2\}$ and three bookstores $B = \{S_0, S_1, S_2\}$. Each region is labeled by a pair of generating points; for example, (R_i, S_j) represents the Voronoi region $V(A, B; R_i, S_j)$.

We see that any pair (R_i, S_j) for $i, j = 0, 1, 2$, $V(A, B; R_i, S_j)$ has a nonempty region in this particular diagram. In general, however, some pairs of restaurants and bookstores may not have nonempty regions.

3   Basic Properties

In this section, we consider basic properties of the round-tour Voronoi diagram defined in the last section.

Property 1. Let $a \in A$ and $b \in B$ and s be a positive real number satisfying $s > 2d(a, b)$. The trajectory of the point z satisfying $l_z(a, b) = s$ forms an ellipse.

Proof. The condition $l_z(a, b) = s$ can be expressed as

$$d(a, p) + d(b, p) = s - d(a, b). \tag{4}$$

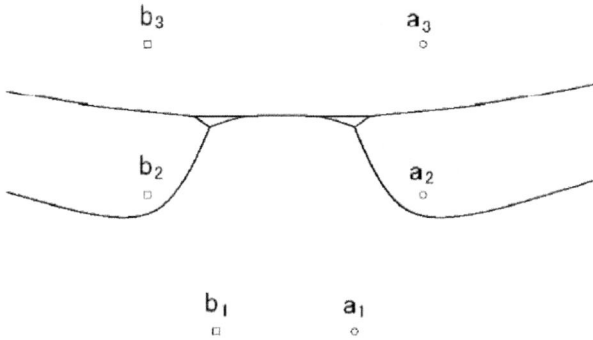

Fig. 3. Round-tour Voronoi diagram with a disconnected region

Because s and $d(a, b)$ are constants, Eq. (4) means that the sum of the distances of p from a and b is constant. Hence, p moves on an ellipse with foci a and b.

Property 2. Let $A = \{a\}$ and $B = \{b, c\}$. Then the boundary between $V(A, B; a, b)$ and $V(A, B; a, c)$ is one branch of the hyperbola with foci b and c.

Proof. The boundary is $\{z \in \mathbb{R}^2 \mid l_z(a, b) = l_z(a, c)\}$. We see

$$\{z \mid l_z(a, b) = l_z(a, c)\}$$
$$= \{z \mid d(z, a) + d(a, b) + d(b, z) = d(z, a) + d(a, c) + d(c, z)\}$$
$$= \{z \mid d(a, b) + d(b, z) = d(a, c) + d(c, z)\}$$
$$= \{z \mid d(b, z) - d(c, z) = d(a, c) - d(a, b)\}.$$

Because $d(a, c) - d(a, b)$ is a fixed constant, the difference of the distances from z to b and c is constant, which means that the boundary point z moves on the hyperbola with foci b and c.

Property 3. A Voronoi region of the round-tour Voronoi diagram is not necessarily connected.

This property can be shown by an example. Consider the first generator set $A = \{a_1, a_2, a_3\}$ where

$$a_1 = (10, -20), \ a_2 = (20, 0), \ a_3 = (20, 22),$$

and the second generator set $B = \{b_1, b_2, b_3\}$ where

$$b_1 = (-10, -20), \ b_2 = (-20, 0), \ b_3 = (-20, 22).$$

The round-tour Voronoi diagram for A and B is shown in Figure 3. In this figure, there are two small triangle-like regions. These two regions together constitute the Voronoi region $V(A, B; a_2, b_2)$. Thus, the Voronoi region is not necessarily connected.

For point $z \in \mathbb{R}^2$ and positive real ϵ, let $U(z, \epsilon)$ be the set of all points that are within the distance ϵ from z. We call $U(z, \epsilon)$ the ϵ-*neighbor* of point z.

Property 4. Let $a \in A, b \in B, z \in \mathbb{R}^2$, and ϵ be a positive real number. For any $z' \in U(z, \epsilon)$, the following inequality is satisfied.

$$l_{z'}(a, b) \leq l_z(a, b) + 2\epsilon \tag{5}$$

Proof. Suppose that we are at point z'. We can visit both points a and b by a round tour visiting z, a, b, z in this order and returning to z'. The length of this round tour is

$$d(z', z) + l_z(a, b) + d(z, z'). \tag{6}$$

The shortest round tour for z' is not longer than this tour, and hence we get

$$l_{z'}(a, b) \leq l_z(a, b) + 2d(z', z)$$
$$\leq l_z(a, b) + 2\epsilon. \tag{7}$$

Property 5. For any $x, x' \in A$ and $y, y' \in B$, if

$$d(x, y) > d(x', y) + d(x, y'), \tag{8}$$

then $V(A, B; x, y)$ is empty.

Proof. First, suppose that

$$d(z, x') \geq d(z, y'). \tag{9}$$

We then obtain

$$
\begin{aligned}
& l_z(x, y) \\
= \ & d(z, x) + d(x, y) + d(y, z) \\
> \ & d(z, x) + d(x', y) + d(x, y') + d(y, z) \\
= \ & d(z, x) + d(x, y') + d(x', y) + d(y, z) \\
& \text{(because of Ineq.(8))} \\
\geq \ & d(z, x) + d(x, y') + d(x', z) \\
& \text{(because of the triangular inequality)} \\
\geq \ & d(z, x) + d(x, y') + d(z, y') \\
& \text{(because of Ineq.(9))} \\
= \ & l_z(x, y').
\end{aligned}
\tag{10}
$$

Hence, we get

$$l_z(x, y) > l_z(x, y'). \tag{11}$$

Secondly, suppose that

$$d(z, x') \leq d(z, y'). \tag{12}$$

Then, by a symmetric argument, we obtain

$$l_z(x, y) \geq l_z(x', y). \tag{13}$$

From Ineqs. (11) and (13), we obtain Property 5.

Property 6. If $a \in V(B; b)$ or $b \in V(A; a)$, then $V(A, B; a, b)$ is nonempty.

Proof. Suppose that $a \in V(B; b)$. Then for any $b' \in B$, we get

$$d(a, b) \le d(a, b'). \tag{14}$$

This is equivalent to

$$d(a, a) + d(a, b) + d(b, a) \le d(a, a) + d(a, b') + d(b', a). \tag{15}$$

Therefore, we get

$$l_a(a, b) \le l_a(a, b'). \tag{16}$$

On the other hand, we get

$$\begin{aligned}
& l_a(a, b') \\
&= d(a, a) + d(a, b') + d(b', a) \\
&= d(a, b') + d(b', a) \\
&\le d(a, a') + d(a', b') + d(b', a) \\
&\quad \text{(because of the triangular inequality)} \\
&= l_a(a', b').
\end{aligned} \tag{17}$$

Combining Ineqs. (16) and (17), we get

$$l_a(a, b) \le l_a(a', b') \tag{18}$$

for any $a' \in A$ and $b' \in B$. This implies $a \in V(A, B; a, b)$, and hence $V(A, B; a, b) \neq \emptyset$.

Next, we suppose $b \in V(A; a)$. Then, by a symmetric argument, we get $b \in V(A, B; a, b)$ and have $V(A, B; a, b) \neq \emptyset$. This completes the proof.

For any finite set X, let $|X|$ denote the number of elements of X.

Property 7. Let $|A| = m$ and $|B| = n$. Then, the number of nonempty Voronoi regions of the round-tour Voronoi diagram can be as small as $\max(m, n)$.

Proof. We prove this property by giving an example of the Voronoi diagram. Without losing generality, we assume that $m \le n$. We consider generator sets $A = \{a_1, \ldots, a_m\}$ and $B = \{b_1, \ldots, b_n\}$ such that

$$\begin{aligned}
a_i = b_i &= (i, 0) \text{ for } i = 1, \ldots, m, \\
b_i &= (i, 0) \quad \text{for } i = m + 1, \ldots, n
\end{aligned}$$

as shown in Figure 4. We show that this diagram has only n regions. First we prove that for any $a \in A$ and $b \in B$, if $V(A, B; a, b)$ is not empty, then b is in $V(A; a)$ in this setting.

Let z be an arbitrary point and a be the point in A that is the nearest to z. In this situation, going straight from z to a and returning to z gives the shortest

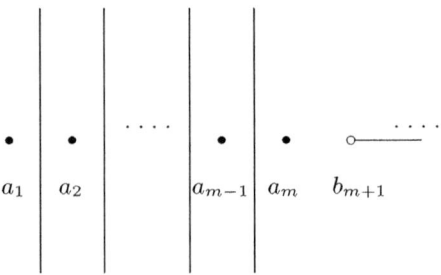

Fig. 4. Round-tour Voronoi diagram with n regions

round tour because we also visit a point in B at point a. We have $a \in V(A;a)$ because a is the nearest point in A to z. Moreover, we have $a \in V(A;a)$ and $V(A;a)$ is convex. Hence, any point in the shortest round tour is in $V(A;a)$. Consequently, we get for any $a \in A$ and $b \in B$, if $V(A,B;a,b)$ is not empty, then b is in $V(A;a)$. Checking the distances, we find that for $i = 1, 2, \ldots, m$, b_i is only in $V(A;a_i)$, and that for $i = m+1, \ldots, n$, b_i is only in $V(A;a_m)$. This completes the proof.

Note that some of the Voronoi regions can not have positive area although they are nonempty. An example of such a region is the Voronoi region $V(A,B;a_m, b_{m+1})$ in Figure 4. The region $V(A,B;a_m,b_{m+1})$ forms the half line starting at b_{m+1} in the positive direction of the x axis. In fact, any point on this half line has the shortest round tour visiting a_m and b_{m+1}. However, any point near to, but not on, this half line can visit a_m and b_m in a round tour shorter than that visiting a_m and b_{m+1}. Thus, the region $V(A,B;a_m,b_{m+1})$ has no area. Let us call the Voronoi region with positive area a *proper region*.

Property 8. Let $|A| = m$ and $|B| = n$. Then, the number of proper Voronoi regions of the round-tour Voronoi diagram can be as small as 1.

Proof. We prove the property by giving an example. Suppose that $A = \{a_1, \ldots, a_m\}$ and $B = \{b_1, \ldots, b_n\}$ such that

Let p be an arbitrary point outside the x axis. Then, as shown in Figure 5, the triangle with the vertices p, a_1, and b_1, shown by broken lines has the smallest perimeter among all triangles with the vertices p, a_i, and b_j for $i = 1, 2, \ldots, m$ and $j = 1, 2, \ldots, n$. Hence, any point outside the x axis belongs to the Voronoi region of the pair (a_1, b_1). Consequently, the pair (a_1, b_1) only has a positive area.

Property 9. Let $|A| = m$ and $|B| = n$. Then the number of nonempty Voronoi regions of the round-tour Voronoi diagram can be as large as mn.

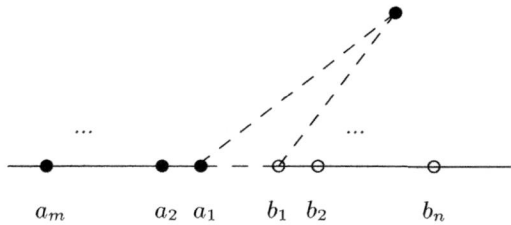

Fig. 5. Round-tour Voronoi diagram with only one positive-area region

Proof. We prove this property by giving an example of a Voronoi diagram. Let

$$A = \{(s_i \cos 60°, s_i \sin 60°) \mid s_i \in [1, 1.5], i = 1, 2, \ldots, m\},$$
$$B = \{(t_j, 0) \mid t_j \in [1, 1.5], j = 1, 2, \ldots, n\}.$$

Suppose that $a \in A$ and $b \in B$ are any pair of generators. We will show that there exists a point $c \in \mathbb{R}^2$ such that $c \in V(A, B; a, b)$. Let L be the line passing through a and b. As shown in Figure 6, let L_1 be the line that is a mirror image of L with respect to the line $y = 2x$, and let L_2 be the mirror image of L with respect to the line $y = 0$. Let c be the point of intersection of L_1 and L_2. Finally, let c_A and c_B be the mirror images of c with respect to $y = 2x$ and $y = 0$ respectively. Note that both c_A and c_B are on the line L, but neither c_A nor c_B is on the line segment connecting a and b. For any $a' \in A$ and $b' \in B$, we get

$$d(c, a') + d(a', b') + d(b', c)$$
$$= d(c_A, a') + d(a', b') + d(b', c_B)$$
$$\geq d(c_A, c_B)$$

because the points c_A, a, b and c_B are on the line L in this order. This implies that $c \in V(A, B; a, b)$. Thus, for any pair $a \in A$ and $b \in B$, the Voronoi region $V(A, B; a, b)$ is nonempty. Consequently, the number of Voronoi regions can be as large as mn.

Figure 7 shows an example of a round-tour Voronoi diagram with $O(n^2)$ Voronoi regions for $|A| = |B| = n$.

Hanniel and Barequet(2009) study a slightly different version of the Voronoi diagram called the triangle-perimeter two-site Voronoi diagram, in which all generators belong to the same type. They proved that their version of the Voronoi diagram has only $O(n)$ non-empty Voronoi regions. Our result, Property 9, shows a remarkable difference between the round-tour Voronoi diagram and the triangle-perimeter Voronoi diagram.

Property 10. The boundary lines of the round-tour Voronoi diagram are piecewise polynomials of degrees upto 16.

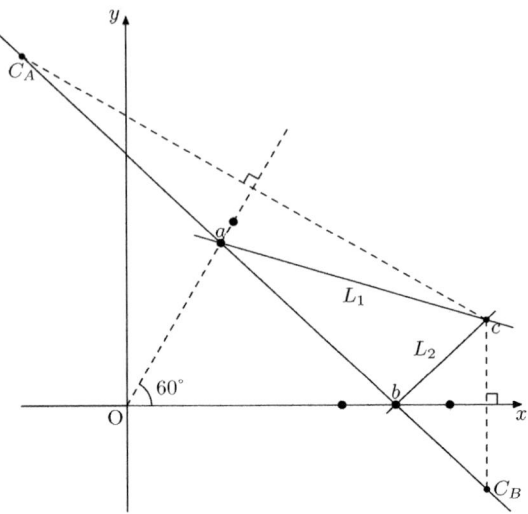

Fig. 6. Three edges of a triangle moved to a common line through two mirrors

Proof. Let the coordinates of four points be $a_1 = (x_1, y_1)$, $b_2 = (x_2, y_2)$, $a_3 = (x_3, y_3)$, and $b_4 = (x_4, y_4)$. Then, we get

$$l_z(a_1, b_2) = \sqrt{(x - x_1)^2 + (y - y_1)^2} + \sqrt{(x_1 - x_2)^2 + (y_1 - y_2)^2}$$
$$+ \sqrt{(x - x_2)^2 + (y - y_2)^2}, \tag{19}$$

$$l_z(a_3, b_4) = \sqrt{(x - x_3)^2 + (y - y_3)^2} + \sqrt{(x_3 - x_4)^2 + (y_3 - y_4)^2}$$
$$+ \sqrt{(x - x_4)^2 + (y - y_4)^2}. \tag{20}$$

Then, the common boundary of the Voronoi region of (a_1, b_2) and that of (a_3, b_4) is represented by the equation:

$$l_z(a_1, b_2) = l_z(a_3, b_4). \tag{21}$$

Let C denote the value of the left-hand side (and consequently the right-hand side) of this equation. Then, we can rewrite eq. (19) as

$$\sqrt{(x - x_1)^2 + (y - y_1)^2} + \sqrt{(x - x_2)^2 + (y - y_2)^2}$$
$$= C - \sqrt{(x_1 - x_2)^2 + (y_1 - y_2)^2}. \tag{22}$$

Squaring the both sides of this equation, we obtain

$$2\sqrt{(x - x_1)^2 + (y - y_1)^2}\sqrt{(x - x_2)^2 + (y - y_2)^2}$$
$$= (C - \sqrt{(x_1 - x_2)^2 + (y_1 - y_2)^2})^2$$
$$- (x - x_1)^2 - (y - y_1)^2 - (x - x_2)^2 - (y - y_2)^2 \tag{23}$$

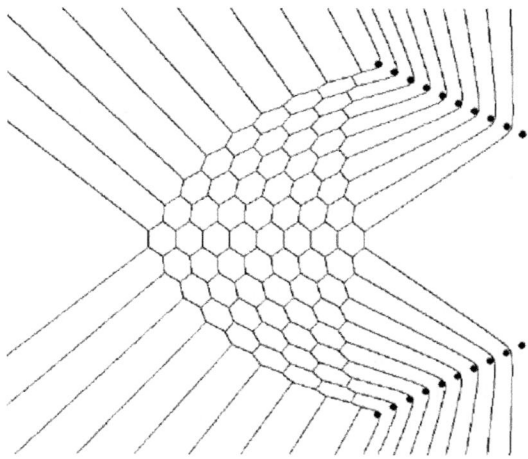

Fig. 7. Round-tour Voronoi diagram with $O(n^2)$ Voronoi regions

and squaring them again, we get dgree-4 polynomial equation in x and y, which we write as

$$f_0 x^4 + f_1 x^3 + f_2 x^2 + f_3 x + f_4 = 0, \qquad (24)$$

where f_i is a degree-i polynomial in y. Similarly from eq. (20), we get another degree-4 polynomial equation, which we write as

$$g_0 x^4 + g_1 x^3 + g_2 x^2 + g_3 x + g_4 = 0, \qquad (25)$$

where g_i is a degree-i polynomial in y.

In order to eliminate x from eqs. (24) and (25), we can apply the Sylvester's elimination method ([11]). That is, the result of eliminating x from eqs. (24) and (25) is represented by the equation:

$$\begin{vmatrix} f_0 & f_1 & \cdots & f_4 & & & \\ & f_0 & f_1 & \cdots & f_4 & & \\ & & \cdots & & & & \\ & & & f_0 & f_1 & \cdots & f_4 \\ g_0 & g_1 & \cdots & g_4 & & & \\ & g_0 & g_1 & \cdots & g_4 & & \\ & & \cdots & & & & \\ & & & g_0 & g_1 & \cdots & g_4 \end{vmatrix} = 0. \qquad (26)$$

Since f_i and g_i are degree-i polynomials in y, this equation is the degree-16 polynomial equation.

4   Algorithm for a Digital-Picture Approximation of the Round-Tour Voronoi Diagram

The boundary curves of the round-tour Voronoi diagram are very complicated in general, and hence it is not easy to construct an exact diagram in a short time.

Indeed, the boundary lines are high-degree polynomials, as shown by Property 10. Hence, we propose an algorithm for constructing a digital-picture approximation of the diagram.

We consider grid points $p_{ij} = (i, j)$ with integer coordinates for $i = 1, 2, \ldots, M$ and $j = 1, 2, \ldots, N$.

For each p_{ij}, there exists a pair (a, b), where $a \in A$ and $b \in B$, such that $p_{ij} \in V(A, B; a, b)$.

We define $D(i, j)$ as

$$D(i, j) = (a, b) \text{ if } p_{i,j} \in V(A, B; a, b).$$

If such a pair (a, b) is not unique, we assign the lexicographically smallest pair (a, b) to $D(i, j)$ for prespecified orders of generators in A and B. Thus, $D(i, j)$ where $i = 1, \ldots, M$ and $j = 1, \ldots, N$ are defined uniquely. We call the set of assignments

$$\{D(i, j) \mid i = 1, \ldots, M, j = 1, \ldots, N\}$$

the *digital-picture approximation of the round-tour Voronoi diagram*, or the *digital Voronoi diagram* for short. The digital Voronoi diagram can be constructed straightforwardly if we do not care about the efficiency; that is, for each grid point p_{ij}, we compute the lengths $l_{p_{ij}}(a, b)$ of the round tour for all pairs (a, b) and assign the one that realizes the minimum of $D(i, j)$. We call this naive algorithm *Algorithm 1*.

Algorithm 1. [naive method] For each grid point p_{ij}, we compute the lengths $l_{p_{ij}}(a, b)$ of the round tours for all $a \in A$ and $b \in B$. We assign the pair that attains the minimum of length $D(i, j)$.

Suppose that we know $l_{p_{ij}}(a, b)$ for some $a \in A$ and $b \in B$. Then, the pair (a', b'), where $a \in A$ and $b \in B$, cannot attain the minimum round tour at p_{ij} if

$$d(p_{ij}, a') > l_{p_{ij}}(a, b). \tag{27}$$

This property can be used to prune the points that cannot attain the minimum round tour. Using this property, we consider the following algorithm.

Algorithm 2. For each grid point p_{ij}, we first compute the length of the round tour under the assumption that p_{ij} belongs to the same Voronoi region as its neighbor does. We then prune the generators a with $d(p_{ij}, a)$ greater than the length of the round tour, and finally compute the true Voronoi region using only the remaining pairs of generators.

We experimentally compared the processing times of Algorithms 1 and 2 for various pairs of generators. The results are summarized in Figure 8, where the horizontal axis represents the number of generators of each kind on a linear scale, and the vertical axis represents the computation time on a logarithmic scale.

From this figure, we can see that the pruning strategy adopted in Algorithm 2 is very effective in constructing the digital Voronoi diagram.

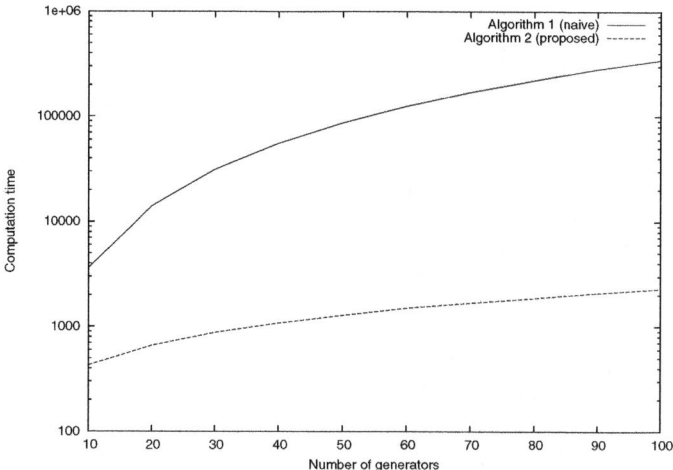

Fig. 8. Computation time for constructing digital round-tour Voronoi diagrams

5   Concluding Remarks

We have proposed a new generalization of the Voronoi diagram named the round-tour Voronoi diagram. This Voronoi diagram is the partition of the plane according to what pair (a, b) of two different kinds of generators attains the shortest round tour. We have studied basic properties of this Voronoi diagram and constructed an algorithm for computing a digital-picture approximation of the diagram.

This generalization is motivated by facility location analysis considering the interaction between two different kinds of facilities such as restaurants and bookstores. Our next task is to apply the round-tour Voronoi diagram to such facility location analysis.

In this paper, we considered two different sets of generators. This can be further generalized to three or more different kinds that a customer wants to visit in the shortest round tour. This direction of the generalization is more complicated because the order of the generators that the customer visits is also important. Full analysis of this approach will be a future work.

Another work for the future is to construct a method for computing the exact Voronoi diagram instead of the digital-picture approximation. Since the boundary curves are represented by high-degree polynomial equations, it is also important to make the algorithm robust against numerical errors.

Acknowledgments

This work is supported by the Grant-in-Aid for Basic Scientific Research (B) No. 20360044 and (B) No. 20300098 from the Japan Society for the Promotion of Science.

References

1. Aronov, B.: On the geodesic Voronoi diagram of point sites in a simple polygon. In: Proceedings of the ACM Symposium on Computational Geometry, Waterloo, Ontario, pp. 39–43 (1989)
2. Aronov, B., O'Rourke, J.: Nonoverlap of the star unfolding. Discrete and Computational Geometry 8, 219–250 (1992)
3. Ash, P.F., Bolker, E.D.: Generalized Dirichlet tessellations. Geometriae Dedicata 20, 209–243 (1986)
4. Aurenhammer, F.: Power diagrams — Properties, algorithms and applications. SIAM Journal of Computing 16, 78–96 (1987)
5. Barequet, G., Dickerson, M.T., Drysdale, R.L.S.: 2-point site Voronoi diagrams. Discrete Applied Mathematics 122, 37–54 (2002)
6. Carter, G.M., Chaiken, J.M., Ignall, E.: Response areas for two emergency units. Operations Research 20, 571–594 (1972)
7. Fujii, H.: Round-Tour Voronoi Diagrams and their Applications; Thesis submitted for Bachelor's degree, Department of Mathematical Engineering and Information Physics, University of Tokyo (2008) (in Japanese)
8. Hakimi, S.L., Labbe, M., Schmeichel, E.: The Voronoi partition of a network and its implications in location theory. ORSJ Journal on Computing 4, 412–417 (1992)
9. Hanniel, I., Barequet, G.: On the triangle-perimeter two-side Voronoi diagram. In: Proceedings of the Sixth International Symposium on Voronoi Diagrams in Science and Engineering, Denmark, June 2009, pp. 129–136 (2009)
10. Imai, H., Iri, M., Murota, K.: Voronoi diagram in the Laguerre geometry and its applications. SIAM Journal of Computing 14, 93–105 (1985)
11. Matematical Society of Japan (ed.): Sugaku Jiten (Dictionary of Mathematics), 3rd edn. Iwanami Shoten, Tokyo (1985) (in Japanese)
12. Nishida, T., Sugihara, K.: Boat-sail Voronoi diagram and its computation based on a cone approximation scheme. Japan Journal of Industrial and Applied Mathematics 22, 367–383 (2005)
13. Ohyama, T.: Some Voronoi diagrams that consider consumer behavior. In: Proceedings of the International Symposium on Voronoi Diagrams in Science and Engineering, Tokyo, September 2004, pp. 191–202 (2004)
14. Ohyama, T.: Application of the additively weighted Voronoi diagram to flow analysis. In: Proceedings of the 2nd International Symposium on Voronoi Diagrams in Science and Engineering, Seoul, October 2005, pp. 22–32 (2005)
15. Okabe, A., Boots, B., Sugihara, K., Chiu, S.N.: Spatial Tessellations — Concepts and Applications of Voronoi Diagrams, 2nd edn. John Wiley, Chichester (2000)
16. Shamos, M.I., Hoey, D.: Closest-point problems. In: Proceedings of the 16th Annual IEEE Symposium on Foundations of Computer Science, pp. 151–162 (1975)
17. Sugihara, K.: Approximation of generalized Voronoi diagrams by ordinary Voronoi diagrams. In: Computer Vision, Graphics, and Image Processing: Graphical Models and Image Processing, vol. 55, pp. 522–531 (1993)
18. Sugihara, K.: Three-dimensional convex hull as a fruitful source of diagrams. Theoretical Computer Science 235, 325–337 (2000)

Protein-Ligand Docking Based on Beta-Shape

Chong-Min Kim[1], Chung-In Won[1], Jae-Kwan Kim[1],
Joonghyun Ryu[1], Jong Bhak[2], and Deok-Soo Kim[1],[*]

[1] Department of Industrial Engineering, Hanyang University, 17 Haengdang-dong,
Seongdong-gu, Seoul, 133-791, South Korea
Tel.: +82-2-2220-0472; Fax: +82-2-2292-0472
{cmkim,wonci,jkkim,jhryu}@voronoi.hanyang.ac.kr, dskim@hanyang.ac.kr
[2] Theragen Inc. Shiheung, Gyeonggi-do, 429-450, South Korea
jongbhak@yahoo.com

Abstract. Protein-ligand docking is to predict the location and orientation of a ligand with respect to a protein within its binding site, and has been known to be essential for the development of new drugs. The protein-ligand docking problem is usually formulated as an energy minimization problem to identify the docked conformation of the ligand. A ligand usually docks around a depressed region, called a pocket, on the surface of a protein. Presented in this paper is a docking method, called BetaDock, based on the newly developed geometric construct called the β-shape and the β-complex. To cope with the computational intractability, the global minimum of the potential energy function is searched using the genetic algorithm. The proposed algorithm first locates initial chromosomes at some locations within the pocket recognized according to the local shape of the β-shape. Then, the algorithm proceeds generations by taking advantage of powerful properties of the β-shape to achieve an extremely fast and good solution. We claim that the proposed method is much faster than other popular docking softwares including AutoDock.

Keywords: Voronoi diagram of spheres, quasi-triangulation, beta-shape, beta-complex, BetaDock, docking, assignment problem, optimization.

1 Introduction

Given a protein, called a receptor, and a small molecule, called a ligand, analyzing their interactions is important for understanding the biological function induced by them. The protein-ligand docking problem is to predict the location and orientation of a ligand relative to the binding site of a target protein. A computer-aided docking process is an important task for structure-based drug design as it identifies lead compounds by minimizing the potential energy of a ligand with respect to a receptor.

Docking between a protein and a receptor usually occurs around a depressed region, called a pocket, on the surface of a receptor. The docking problem is computationally difficult because there are many ways of putting the two molecules

---

[*] Corresponding author.

M.L. Gavrilova et al. (Eds.): Trans. on Comput. Sci. IX, LNCS 6290, pp. 123–138, 2010.
© Springer-Verlag Berlin Heidelberg 2010

together and the number of possibilities that must be examined grows exponentially with respect to the size of the molecules. The orientation space of two molecules is also so large as to make exhaustive search methods prohibitive. It is known that the docking problem is NP-hard [31,39]. Hence, a heuristic approach is inevitable.

Most docking softwares are based on heuristics and therefore require some initial solutions for the early stage of the solution search. It seems that using predicted pockets on the surface of a receptor is a promising approach for generating a good initial solution. Considering millions of entries in chemical data bases, manually identifying pockets for docking may be an almost infeasible approach. Therefore, an automatic recognition of pockets and the fast evaluation of the binding between a compound and a pocket are very important in the study of protein-ligand docking for the development of new drugs [26].

Compared to the physicochemical approach, efforts to understand the geometry of molecules have started only very recently [3,9,11,28,35,38]. Contemporary agreement is that the geometry is as important and critical as the physicochemical properties for biological systems. Hence, the research on the geometry in biological systems will provide new challenges as well as opportunities for the community of geometers.

In this paper, we present a docking method and software, called the BetaDock, that is based on the recently announced geometric construct, the β-shape and the β-complex. The β-shape and β-complex have proved their powerful capabilities for the proximity problems among atoms in molecules [15,17,23,36]. The Beta-Dock is much faster than most popular docking softwares including AutoDock. This paper is the revised version of a conference paper presented at the 6th International Symposium on Voronoi Diagrams in Science and Engineering held in Copenhagen, 2009 [14].

The paper is organized as follows: Section 2 introduces related previous works. Section 3 presents the issues related to the topology among the atoms of proteins and provides the definition of the β-shape and β-complex. Section 4 describes the idea of pocket recognition used in this paper. Section 5 defines the docking problem presented in this paper. Section 6 proposes the docking method used in BetaDock. Section 7 presents an experimental result and Section 8 concludes the paper.

2   Previous Works

Since the initial proposal by Kuntz et al. in 1982 [27], several studies have been conducted on the automatic docking simulation of a ligand to a protein. As the *ab initio* approach is usually infeasible from a computational point of view, the docking problem is usually formulated using an empirical model of the energy minimization problem among the atoms in both protein and ligand. Even in the empirical model, the minimization problem has an exponential number of local minima with respect to the problem size and ordinary local search methods in the classical optimization fail to locate the global minima. If the conformational flexibility of a ligand and/or a protein is also reflected in the computation, the solution space becomes even larger. Hence, a heuristic is inevitable and the genetic algorithm (GA) is one of the favored approaches.

For example, [24] describe the use of GA for the docking problem in detail. There are also popular softwares such as AutoDock, DARWIN, DIVALI, GOLD, EADock, FITTED, PSI-DOCK, etc. which simulate the protein-ligand docking. Note that many popular softwares assume the flexibility of ligand and the classical concept of GA is sometimes modified to fit special requirements in each software. For example, Lamarckian GA used in AutoDock or the tabu search-enhanced GA in PSI-DOCK have been reported as exploring the conformational space of ligand more efficiently [32,34].

Most heuristic algorithms for docking require to define initial position(s) of a ligand in the vicinity of a protein at the beginning of the algorithm. The core part of DOCK, the first developed docking software by Kuntz et al. [27], is still being used in some softwares. DOCK first attempts to find binding sites based on some physical properties of a protein. Recent studies suggest to locate the initial positions of ligand around the predicted binding site [7,33].

However, many docking softwares do not use the information about potential binding sites. Instead, they manually locate the initial positions of a ligand around the protein. For example, AutoDock assists users to place the initial chromosomes at random locations around the user-defined bounding box. When the initial solutions are manually created, the accuracy of the result usually depends on the biological knowledge of the individual user of the software. The bounding box around a protein may provide a degree of freedom to the user so that he or she can easily enforce the initial chromosomes around the anticipated binding sites on the protein. This approach implies that the recognition of probable binding sites, which is usually a pocket, may be useful in the docking simulation.

With this observation in mind, several studies have been conducted to automatically extract the pockets from proteins. Geometric approaches to recognize pockets on a protein usually involve the definition of surfaces on a protein. Initial studies on this approach primarily created a 3D spatial lattice of the space occupied by a protein and used simple but computationally inefficient techniques to reason the relative relations among the grids in the lattice to extract the exterior boundary of the protein, and then they recognized the depressed regions on the surface of the protein [8,12,13,41]. [17] provides a mathematically well-defined and computationally efficient algorithm to extract pockets on the *unambiguous* boundary of proteins using the β-shape. In the same paper, the authors also describe a good survey of approaches to the pocket extraction problem. Even though some studies used a similarly unambiguous approach of extracting pockets using α-shape [10,28,35], the α-shape considers that atoms in molecules are mono-sized. Even the weighted α-shape suffers from computational inefficiency compared to the β-shape.

3   Topology Representation for Protein

3.1   Voronoi Diagram of Atoms and Quasi-triangulation

Suppose that $A = \{a_1, a_2, \ldots, a_n\}$ is a molecule where $a_i = (p_i, r_i)$ is a three-dimensional sphere with a center $p_i = (x_i, y_i, z_i)$ and a radius r_i. Let $VD(A)$ denote the Voronoi diagram of the sphere set A. $\mathcal{VD}(A)$ is also frequently referred to as the additively weighted Voronoi diagram [30]. Then, $\mathcal{VD}(A) = (V^{\mathcal{V}}, E^{\mathcal{V}}, F^{\mathcal{V}}, C^{\mathcal{V}})$ where $V^{\mathcal{V}} = \{v_1^{\mathcal{V}}, v_2^{\mathcal{V}}, \ldots\}$, $E^{\mathcal{V}} = \{e_1^{\mathcal{V}}, e_2^{\mathcal{V}}, \ldots\}$, $F^{\mathcal{V}} = \{f_1^{\mathcal{V}}, f_2^{\mathcal{V}}, \ldots\}$ and $C^{\mathcal{V}} = \{c_1^{\mathcal{V}}, c_2^{\mathcal{V}}, \ldots c_n^{\mathcal{V}}\}$ denote the sets of Voronoi vertices, Voronoi edges, Voronoi faces and Voronoi cells (or regions) in the Voronoi diagram, respectively. By definition, a Voronoi vertex $v^{\mathcal{V}}$ is the center of an empty sphere tangent to four nearby balls, and a Voronoi edge $e^{\mathcal{V}}$ is defined as a set of points equi-distant from the boundaries of three nearby balls. A Voronoi face $f^{\mathcal{V}}$ is the connected set of points equidistant to the boundaries of two neighboring balls. Note that more than one Voronoi edge or face can be defined by three or two balls, respectively. A Voronoi cell $c_i^{\mathcal{V}}$ is defined as $c_i^{\mathcal{V}} = \{q \mid \text{dist}(q, p_i) - r_i \leq \text{dist}(q, p_j) - r_j, i \neq j\}$. A Voronoi edge $e^{\mathcal{V}}$ is a conic curve and a Voronoi face $f^{\mathcal{V}}$ is a hyperboloid. The robust and fast computation of such $\mathcal{VD}(A)$ has been a major challenge for several years in computational geometry. Recently, the efficient and robust computation of the Voronoi diagram of spheres has become practical and readers are referred to [5,18,19,20] for more details of the algorithms and their applications.

Given a Voronoi diagram, its topology is usually stored in the dual structure for both efficiency and convenience. For example, the Voronoi diagram of points is stored in the Delaunay triangulation because it is a simplicial complex that can be stored in a compact data structure allowing a convenient traversal.

However, the dual of the Voronoi diagram of atoms, called a *quasi-triangulation*, is not a simplicial complex because there are a few anomalies that violate the condition for a simplicial complex. Kim et al. presents both the definition and the properties of quasi-triangulation and the compact data structure to store it [21]. It is called a quasi-triangulation because the dual of the Voronoi diagram of three-dimensional spheres is mostly simplicial complex but with a few violating cases. In particular, the dual of the Voronoi diagram of protein atoms contains only very few violating cases. If the topology of the Voronoi diagram of spheres is to be directly stored, the radial edge data structure has to be used [6].

The conversion between the Voronoi diagram and the quasi-triangulation takes $O(m)$ time in the worst case where m is the number of geometric entities in either structure. Hence, the Voronoi diagram and the quasi-triangulation are equivalent from both a computational and an informational point of view. For the details of quasi-triangulation, see [21].

3.2 β-Shape and β-Complex

Consider spherical rocks of varying radii scattered within a piece of soft matter. When some subset of the matter is removed by an omni-present spherical eraser with a radius β, the shape left is called the β-hull. The boundary of the β-hull contains the blending surface of the spherical rocks formed by the spherical eraser.

Fig. 1(a) and (b) are the β-hulls of thirteen circular rocks in the plane corresponding to circular disks with different sizes. The spherical eraser is called the β-probe. The molecular surface in biology is indeed equivalent to the β-hull of the molecule, when the spherical rocks correspond to the atoms in a molecule. Let the β-probes for the β-hull in Fig. 1(a) and Fig. 1(b) are denoted as β_1 and β_2 respectively. Then $\beta_1 < \beta_2$.

Given the β-hull of a protein A, we straighten the surface of the β-hull by substituting straight edges for the circular arcs and planar triangles for the spherical triangles where the vertices are the centers of the atoms contributing to the boundary of the β-hull. The straightened object bounded by triangular facets is the β-shape of A and the β-shapes that correspond to β_1 and β_2 are as illustrated in Fig. 1(c) and (d) respectively. The shaded region in the β-shape is the interior of the β-shape.

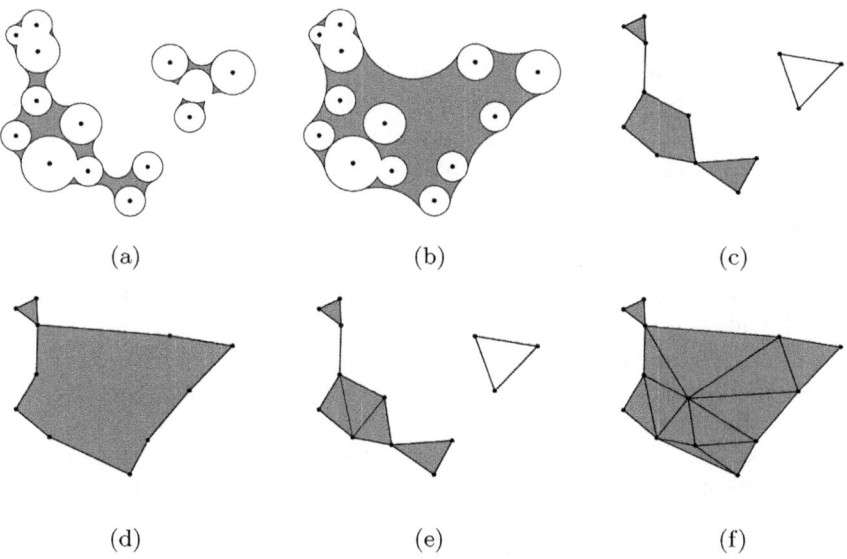

(a)   (b)   (c)

(d)   (e)   (f)

Fig. 1. β-hulls, β-shapes, and β-complexes for the thirteen atoms in $\mathbb{R}^2$: (a) atoms and the β-hull corresponding to a small β-probe with β_1, (b) the β-hull corresponding to a large β-probe with β_2 where $\beta_1 < \beta_2$, (c) and (d) the β-shapes corresponding to β_1 and β_2, and (e) and (f) the corresponding β-complexes, respectively.

The β-complex of an atom set A is a subset of the quasi-triangulation of A. The volume union of the simplex volumes of the β-complex is identical to the volume of the corresponding β-shape. Shown in Fig. 1(e) and (f) are the β-complexes of the β-shapes in Fig. 1(c) and (d), respectively. As shown in the figures, the β-shape is tessellated into a set of simplexes in the β-complex. A simplex σ in the β-complex takes one of the following three *bounding states* [37]. Note that a simplex σ has one of these states if it belongs to a β-complex.

- *singular*: σ belongs to the boundary of the β-shape and does not bound any higher-dimensional simplex in the corresponding β-complex.
- *regular*: σ belongs to the boundary of the β-shape and bounds a higher-dimensional simplex in the corresponding β-complex.
- *interior*: σ does not belong to the boundary of the β-shape, and σ is the intersection between neighboring simplexes of higher dimensions in the corresponding β-complex.

For example in Fig. 1(e), four edges which do not bound any shaded triangle are singular. Two edges that are simultaneously shared by two shaded triangles are interior and the rest edges, that are neither singular nor interior, are regular. All thirteen vertices are regular and all five shaded triangles are interior. Details on the β-shape and β-complex can be found in [23,37].

4   Pocket Extraction via β-Shape

In the course of locating pockets of a protein, it is important to traverse among simplexes efficiently. If the boundary of the β-shape is represented by an efficient data structure such as winged-edge data structure or half-edge data structure, the traversal among simplexes will be very efficient. Unfortunately, however, the β-shape is a non-manifold object in general and has dangling edges as well as dangling faces. The boundary, therefore, cannot be stored directly in data structures for a manifold object. One possible way of solving this problem is to convert a β-shape into a manifold and represent the boundary of the modified β-shape by efficient data structures for manifolds [16,22].

Kim et al. reported an algorithm to extract pockets of a protein using two β-shapes defined by small and large β values [17]. The β-shape defined by a small β value is called an inner β-shape and the one defined by a large β value is called an outer β-shape. Fig. 2 shows examples of the inner and outer β-shapes for a protein in the plane. From this example, we observe that the difference between the two β-shapes can define pockets. For each edge of the outer β-shape, there is zero or one depressed region on the boundary of the protein. When an edge of the outer β-shape coincides with one of the inner β-shape, obviously no pocket is defined.

Although the problem in 3D is not as simple as its 2D counterpart, we make a similar observation. For each face of the outer β-shape of the protein, there can be a corresponding depressed region. A depressed region corresponding to the face of the outer β-shape, however, may not have a clearly defined boundary. Hence, we need to subdivide the boundary of the inner β-shape so that each face of the outer β-shape corresponds to one subdivided region.

Fig. 3 shows the procedure of pocket recognition used in this paper. Shown in Fig. 3(a) is the input protein displayed by its molecular surface. Fig. 3(b)

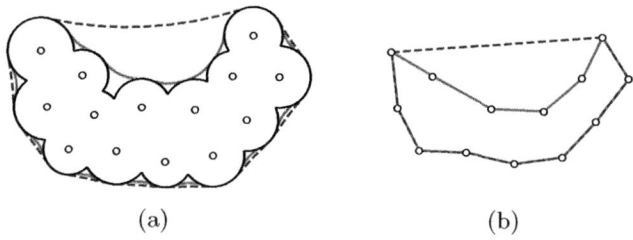

(a)   (b)

Fig. 2. Inner and outer β-shapes (2D analogy)

and (c) are the inner and outer β-shapes, respectively. The two β-shapes are overlapped in Fig. 3(d). The outer β-shape, in this example, is when β is ∞ and therefore it approximates the convex hull of the atom centers. Note that users can determine appropriate β values for inner and outer β-shapes as a parameter of the algorithm. Fig. 3(e) shows the subdivided region on the boundary of the inner β-shape and Fig. 3(f) shows the extracted pocket.

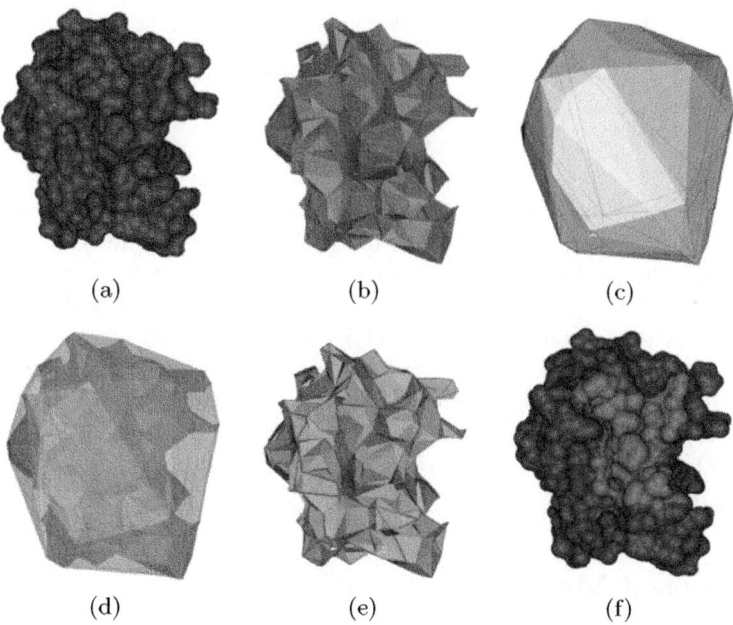

(a)   (b)   (c)

(d)   (e)   (f)

Fig. 3. The protein 1FKG and the corresponding pockets: (a) the receptor shown by the molecular surface, (b) the inner β-shape, (c) the outer β-shape, (d) the overlapped inner and outer β-shapes, (e) the pocket boundary, and (f) the largest pocket (red) on the receptor.

5   Docking Problem

In order to dock a protein and a ligand, the free energy of the system of the protein and the ligand needs to be minimized. Protein-ligand docking is formulated as a high-dimensional non-linear optimization problem associated with an accurate scoring function, which aims to identify the docked conformation of a ligand with the lowest energy.

In our case, both protein receptor and ligand are assumed to be rigid bodies. A protein molecule consists of a set of atoms, each of which is represented by a solid sphere in $\mathbb{R}^3$. Let $R = \{r_1, r_2, \ldots, r_m\}$, and $L = \{l_1, l_2, \ldots, l_n\}$ be a receptor and a ligand, where r_i and l_j are atoms in R and L respectively. Usually, $n << m$. In modeling the forces in the molecular system of a receptor and a ligand, there are six major forces: three forces related among non-bonded atoms and three forces related among bonded atoms. In this paper, we consider the three non-bonded terms only. The forces for the bonded-terms are as follows: the *van der Waals* force, the *electrostatic* force, and the *hydrogen-bond* interactions. Note that this force is in effect at a distance.

Then, the docking problem of R and L can be formulated as Eq. (1) [25].

$$Minimize \sum_{l \in L} \sum_{r \in R} \left\{ \frac{A_{lr}}{d_{lr}^{12}} - \frac{B_{lr}}{d_{lr}^{6}} \right\} + \sum_{l \in L} \sum_{r \in R} 332.0 \frac{q_l q_r}{\varepsilon \cdot d_{lr}}$$
$$+ \sum_{l \in L} \sum_{r \in R} \left\{ \frac{C_{lr}}{d_{lr}^{12}} - \frac{D_{lr}}{d_{lr}^{10}} \right\} \tag{1}$$

$$s.t. \ R \ and \ L \ are \ rigid \ body$$

In Eq. (1), d_{lr} is the Euclidean distance between the centers of atom pair l and r. The first term is the van der Waals interaction, which is the sum of attractive and repulsive forces between each pair of atoms. A_{lr} and B_{lr} are the constants that depend on the types of atoms l and r. The second term is the electrostatic interaction, which depends on the electric charges of both atoms. q_l and q_r are the point charges of l and r respectively, ε is a dielectric constant, and 332.0 is a factor that converts the electrostatic energy into kilocalories per mol. Finally, the third term is the hydrogen-bond interaction between each pair of atoms wehre C_{lr} and D_{lr} are the constants depending on the hydrogen bond between the atoms l and r.

6   Proposed Method for Docking

In BetaDock, the genetic algorithm (GA) is adopted to search the global minimum of the energy function. The essential idea of the genetic algorithm is the evolution of a population consisting of solutions, which are usually referred to as chromosomes, via genetic operations of the mutation and the crossover to reach a final solution by optimizing a predefined fitness function, given in Eq. (1). In the current version of BetaDock, the β-shape is the fundamental construct used to compute the initial population.

6.1   Model of Chromosome

The population consists of N chromosomes. Assuming that a ligand is a rigid body, each chromosome corresponding to a potential docking pose (i.e., the location and the orientation of a ligand with respect to a receptor) is defined by a set of six parameters: three parameters for the location and three parameters for the orientation. We assume that the receptor is fixed in the space and the ligand moves. Hence, the chromosome c is given as follows:

$$c = (\theta_x, \theta_y, \theta_z, \delta_x, \delta_y, \delta_z) \tag{2}$$

which is popular in that it is frequently used in other studies.

6.2   Initial Population

Our scheme is based on the observation that the best position of a ligand is probably in the vicinity of the receptor surface. Hence, the candidates of a binding site can be geometrically identified as pockets that are recognized using the β-shape of the protein. It is known that protein normally exists in a solvent that is usually water. Hence, it is preferred to set the inner β value with the radius of water molecule, 1.42 Å. For outer β value, the diameter of the minimum enclosing sphere for the ligand seems reasonable from our experimental observation.

Let pocket $P = \{a_1, \ldots, a_{m_P}\}$ be a set of m_P atoms, and $\chi = \{b_1, \ldots, b_k\}$ be a set of center points of k balls where each ball is tangent to a triplet of atoms in P. Let $\xi = \{c_1, \ldots, c_m\}$ be a set of center points of m atoms in a ligand. Then, the superposition between χ and ξ makes ligand to be placed into the pocket. The radius of the biggest atom in the ligand is selected for the radius of tangent ball to avoid collisions between the pocket and the ligand as much as possible after the superposition. It turns out that the structure superposition can be done quickly using the singular value decomposition (SVD) [29]. Hence, the initial population of GA requires solving N structure superpositions because we require N chromosomes in the initial solution of the GA process. To do this, we first make N subsets of χ and then superpose ξ and each subset.

6.3   Genetic Operators

BetaDock follows the ordinary GA process with a fitness function given by Eq.(1) for each chromosome. Then, as the generation repeats, the mutation operator randomly changes the value of a gene and the crossover operator exchanges a set of genes from one parent chromosome to another. Selection was performed by the roulette wheel method and the crossover was executed by the two-point rule. After selecting two individuals from the current population as the parent, the two-point rule randomly selected a consecutive subset of each chromosome where both subsets were at identical locations in the parent chromosomes. The three parameters for the rotation was mutated in the range of 0∼360° and the three parameters for the translation was mutated within the distance equivalent to the radius of the minimum enclosing sphere of the ligand.

7 Experimental Results

The experiment was performed on the cluster computer which consists of 118 nodes in the Voronoi Diagram Research Center [2]; Each node has dual cpu's AMD Opteron Dual core 2.2GHz with 2GB RAM; Linux was the running OS in the experiment. For the experiment, we selected ten protein complexes with bounded ligands (available at the RCSB Protein Data Bank (PDB) [1]) that are a subset of the proteins introduced by Baxter [4] in 1998. These proteins did not include the metallic atom(s) because it is known that metallic ions mediate the protein-ligand interactions. The ten proteins are listed in Table 1.

Table 2 summarizes the default setting of BetaDock parameters for docking simulation. For each protein, we extracted five pockets and distributed the same number of chromosomes in each pocket to minimize the possibility of missing a good solution. There were two terminating conditions: First, BetaDock terminates if the a priori set maximum number of generations in the entire simulation is reached; Second, BetaDock terminates if the a priori set maximum number of generations without improving the current best solution is reached.

Table 1. 10 test complexes from PDB

PDB ID	Receptor		Ligand	
	Name	# atoms	Name	# atoms
1 1FKG	FK-506 binding protein	832	SB3	33
2 1STP	Streptavidin	902	BTN	16
3 1AEC	Actinidin	1,649	E64	33
4 1MRG	α-momorcharin	1,924	AND	15
5 1BLH	β-lactamase	1,999	FOS	15
6 1PTV	Tyrosine phosphatase	2,426	TYR	17
7 1C83	Tyrosine phosphatase	2,427	OAI	18
8 1TDB	Thymidylate synthase	2,583	UFP	21
9 1DBJ	Immunoglobulin	3,354	AE2	21
10 1PGP	6-p dehydrogenase	3,654	6PG	17

Table 2. Default parameter settings for BetaDock

Inner β value for pocket	1.42 Å
Outer β value for pocket	diameter of min. enclosing sphere of ligand
Num. of pockets for init. population	5
Size of Population	100
Num. of max. generation	1,000
Rate crossover	0.5
Rate mutation	0.3
Ending condition	200

For the comparison of the BetaDock performance, we selected AutoDock 4.0 [32] which is the most commonly cited docking program in the scientific literature [40]. AutoDock allows users to choose between the Monte Carlo simulated annealing and the Lamarckian genetic algorithm (LGA) as the meta-heuristic to search the global minimum of the energy function. The default parameters of AutoDock are listed in Table 2. Note that AutoDock creates an initial population around the bounding box enclosing the receptor. For a fair comparison, AutoDock was also set for the rigid-body docking.

Table 3. Default parameter settings for AutoDock

Type of search method	LGA
Size of Population	150
Num. of max. energy evaluation	2,500,000
Num. of max. generations	27,000
Rate mutation	0.02
Rate crossover	0.8
Local search iteration	300
LGA runs	10
Grid spacing	0.375

Table 4 shows the computation times of BetaDock for the selected ten protein-ligand complexes. Complexes were sorted in an increasing order of the number of atoms in the protein. The computation time for creating initial population includes the computation time of Voronoi diamgram (the VD column), the β-shape (the BS column), the pocket extraction (the Pocket column), and the superposition (the Superpos. column) processes. We emphasize here that the computation time for VD can be done as the pre-processing because docking is usually done for many ligands against a single protein of interest.

Table 4. Computation time of BetaDock for 10 complexes

No.	PDB ID	# atoms	Initial population					⑥ GA	Total
			① VD	② BS	③ Pocket	④ Superpos.	⑤ Total (①+②+③+④)		(⑤+⑥)
1	1FKG	832	4.27	0.29	0.09	4.37	4.65	47.91	52.56
2	1STP	902	4.11	0.25	0.08	1.11	4.44	12.82	17.26
3	1AEC	1649	8.68	0.59	0.17	4.45	9.44	95.47	104.91
4	1MRG	1924	10.49	0.71	0.32	1.04	11.52	42.10	53.62
5	1BLH	1999	9.76	0.60	0.21	1.03	10.57	27.80	38.37
6	1PTV	2426	11.64	0.73	0.28	1.33	12.65	43.10	55.75
7	1C83	2427	11.72	0.77	0.25	1.48	12.74	91.77	104.51
8	1TDB	2583	12.91	0.79	0.38	1.96	14.08	75.16	89.24
9	1DBJ	3354	16.52	1.04	0.57	1.95	18.13	61.86	79.99
10	1PGP	3654	17.98	1.19	0.78	1.35	19.95	89.86	109.81

* time: *sec.*

Table 5. Result of docking for 10 complexes by using BetaDock and AutoDock

No.	PDB ID	BetaDock				AutoDock			
		Comp. time	Best gen.	Energy	RMSD	Comp.time	Best gen.	Energy	RMSD
1	1FKG	52.56	53	-16.04	9.50	443.22	2742	371.12	5.45
2	1STP	17.26	8	-12.75	14.42	349.50	2664	62.61	5.52
3	1AEC	104.91	19	-12.61	12.30	505.10	2926	238.75	13.82
4	1MRG	53.62	1	-16.51	13.95	297.80	2299	-4.91	7.38
5	1BLH	38.37	5	-13.72	3.26	349.67	2620	15.36	11.49
6	1PTV	55.75	22	-13.72	16.59	355.67	2610	149.16	18.97
7	1C83	104.51	1	-18.81	2.01	359.38	2700	95.25	17.26
8	1TDB	89.24	63	-13.25	15.40	411.35	2577	-2.13	7.53
9	1DBJ	79.99	2	-14.31	8.34	365.92	2221	-5.29	31.48
10	1PGP	109.81	69	-13.38	27.84	363.73	2683	114.72	13.88

* time: *sec.*, energy: *kcal/mol*, rmsd:$\mathring{A}$

Table 5 compares BetaDock against AutoDock. To conclude, BetaDock found solutions of a quality comparable with that of AutoDock much faster. For both softwares, *Best gen.* column denotes the generation that produced the best solution until the simulation was terminated. In particular, the best solutions for BetaDock were found in the very early generation of the GA process compared to AutoDock. We believe this is the case because the initial populations of Beta-Dock were computed very close to the global optimum because they are already located in good places in the pocket. The comparison of computation time for BetaDock and AutoDock is plotted in Fig. 4. From Fig. 4, it can be seen that the BetaDock is much faster than the AutoDock Note that the computation time for AutoDock includes the energy-grid calculation time for preprocessing.

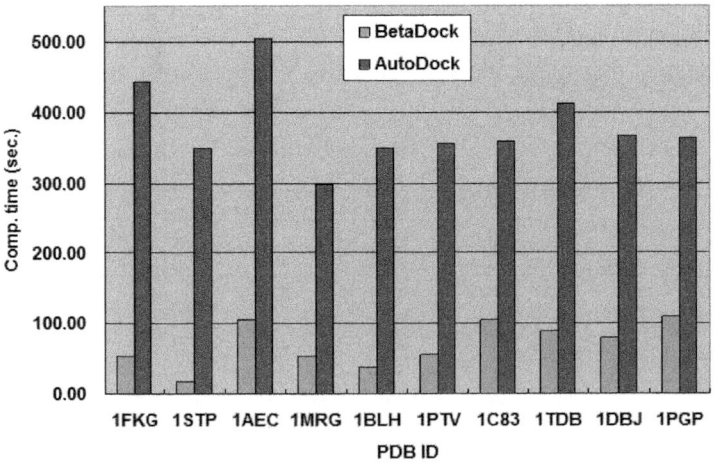

Fig. 4. Comparison of computation time for BetaDock and AutoDock

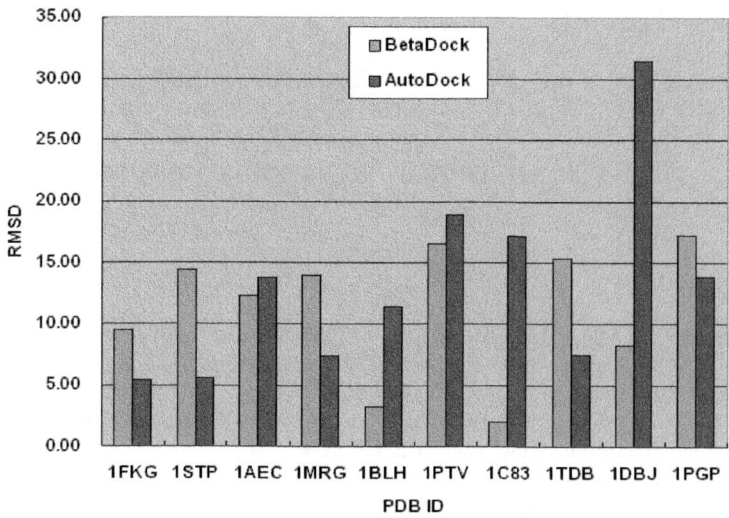

Fig. 5. Comparison of RMSD for BetaDock and AutoDock

The *Energy* column of Table 5 shows the potential energy (fitness value) of the best solution in the unit of *kcal/mol*. The *RMSD* column denotes the *root mean squared deviation* between the corresponding atoms in both the bound ligand and the simulated ligand and has been used as the major criterion for judging the docking results. Hence, the lower RMDS value is, the better the algorithm is. See Fig. 5 to verify that BetaDock is better than AutoDock.

8   Conclusions

BetaDock, the docking method and software proposed in this paper, is based on the theory of the β-shape and the β-complex. We first compute pockets of receptor and locate initial solutions of the genetic algorithm within these pockets using the superposition technique of two point sets. After the initial chromosomes are located within the pockets, the genetic algorithm is applied to search the global optimum using the mutation and the cross-over operators where both are related with the shape of the ligand and the receptor. It turns out that the proposed docking method produces a good solution faster than other commercial docking softwares.

The idea of the docking proposed in this paper needs to be explored in various directions. Particularly, the use of β-shape and the β-complex seems critical because they represent the proximity among related atoms in receptor in a concise form. We also anticipate that different, better approaches for pocket recognition may be possible so that better global solutions can be found faster than the current version of BetaDock.

Acknowledgments

This research was supported by the Korea Science and Engineering Foundation (KOSEF) through the National Research Lab. funded by the Ministry of Science and Technology (No. R0A-2007-000-20048-0) and the Korea Research Foundation Grant (MOEHRD, Basic Research Promotion Fund) (KRF- 2007-314-D00311).

References

1. RCSB Protein Data Bank Homepage (2009), http://www.rcsb.org/pdb/
2. Voronoi Diagram Research Center (2009), http://voronoi.hanyang.ac.kr/
3. Agarwal, P.K., Edelsbrunner, H., Harer, J., Wang, Y.: Extreme elevation on a 2-manifold. In: Proceedings of the 20th Annual ACM Symposium on Computational Geometry, Brooklyn, New York, USA, pp. 357–365 (2004)
4. Baxter, C.A., Murray, C.W., Clark, D.E., Westhead, D.R., Eldridge, M.D.: Flexible docking using tabu search and an empirical estimate of binding affinity. Proteins: Structure, Function, and Genetics 33, 367–382 (1998)
5. Boissonnat, J.-D., Delage, C.: Convex hull and Voronoi diagram of additively weighted points. In: Brodal, G., Leonardi, S. (eds.) ESA 2005. LNCS, vol. 3669, pp. 367–378. Springer, Heidelberg (2005)
6. Cho, Y., Kim, D., Kim, D.-S.: Topology representation for the Voronoi diagram of 3D spheres. International Journal of CAD/CAM 5(1), 59–68 (2005), http://www.ijcc.org
7. Corbeil, C.R., Englebienne, P., Moitessier, N.: Docking ligands into flexible and solvated macromolecules. 1. development and validation of FITTED 1.0. Journal of Chemical Information and Modeling 47, 435–449 (2007)
8. Delaney, J.S.: Finding and filling protein cavities using cellular logic operations. Journal of Molecular Graphics 10, 174–177 (1992)
9. Edelsbrunner, H., Facello, M., Liang, J.: On the definition and the construction of pockets in macromolecules. Discrete Applied Mathematics 88, 83–102 (1998)
10. Edelsbrunner, H., Mücke, E.P.: Three-dimensional alpha shapes. ACM Transactions on Graphics 13(1), 43–72 (1994)
11. Heifets, A., Eisenstein, M.: Effect of local shape modifications of molecular surfaces on rigid-body protein-protein docking. Protein Engineering 16(3), 179–185 (2003)
12. Hendlich, M., Rippmann, F., Barnickel, G.: LIGSITE: Automatic and efficient detection of potential small molecule-binding sites in proteins. Journal of Molecular Graphics & Modeling 15(6), 359–363 (1997)
13. Ho, C.M., Marshall, G.R.: Cavity search: an algorithm for the isolation and display of cavity-like binding regions. Journal of Computer-Aided Molecular Design 4, 337–354 (1990)
14. Kim, C.-M., Won, C.-I., Ryu, J., Bhak, J., Kim, D.-S.: Protein-ligand docking based on β-shape. In: Proceeding of the 6th International Symposium on Voronoi Diagrams in Science and Engineering, pp. 245–253 (2009)
15. Kim, D., Cho, C.-H., Cho, Y., Ryu, J., Bhak, J., Kim, D.-S.: Pocket extraction on proteins via the Voronoi diagram of spheres. Journal of Molecular Graphics and Modelling 26(7), 1104–1112 (2008)
16. Kim, D., Lee, C., Cho, Y., Kim, D.-S.: Manifoldization of β-shapes by topology operators. In: Chen, F., Jüttler, B. (eds.) GMP 2008. LNCS, vol. 4975, pp. 505–511. Springer, Heidelberg (2008)

17. Kim, D.-S., Cho, C.-H., Kim, D., Cho, Y.: Recognition of docking sites on a protein using β-shape based on Voronoi diagram of atoms. Computer-Aided Design 38(5), 431–443 (2006)
18. Kim, D.-S., Cho, Y., Kim, D.: Edge-tracing algorithm for Euclidean Voronoi diagram of 3D spheres. In: Proceedings of the 16th Canadian Conference on Computational Geometry, pp. 176–179 (2004)
19. Kim, D.-S., Cho, Y., Kim, D.: Euclidean Voronoi diagram of 3D balls and its computation via tracing edges. Computer-Aided Design 37(13), 1412–1424 (2005)
20. Kim, D.-S., Cho, Y., Kim, D., Kim, S., Bhak, J., Lee, S.-H.: Euclidean Voronoi diagrams of 3D spheres and applications to protein structure analysis. Japan Journal of Industrial and Applied Mathematics 22(2), 251–265 (2005)
21. Kim, D.-S., Kim, D., Cho, Y., Sugihara, K.: Quasi-triangulation and interworld data structure in three dimensions. Computer-Aided Design 38(7), 808–819 (2006)
22. Kim, D.-S., Lee, C., Cho, Y., Kim, D.: Manifoldization of β-shapes on $o(n)$ time. Computer-Aided Design 42(4), 322–339 (2010)
23. Kim, D.-S., Seo, J., Kim, D., Ryu, J., Cho, C.-H.: Three-dimensional beta shapes. Computer-Aided Design 38(11), 1179–1191 (2006)
24. Kitchen, D.B., Decornez, H., Furr, J.R., Bajorath, J.: Docking and scoring in virtual screening for drug discovery: Methods and applications. Nature 3, 935–949 (2004)
25. Kollman, P.: Free energy calculations: Applications to chemical and biochemical phenomena. Chemical Reviews 93, 2395–2417 (1993)
26. Kuntz, I.D.: Structure-based strategies for drug design and discovery. Science 257(21), 1078–1082 (1992)
27. Kuntz, I.D., Blaney, F.M., Oatley, S.J.: A geometric approach to macromolecule-ligand interactions. Journal of Molecular Biology 161, 269–288 (1982)
28. Liang, J., Edelsbrunner, H., Woodward, C.: Anatomy of protein pockets and cavities: Measurement of binding site geometry and implications for ligand design. Protein Science 7(9), 1884–1897 (1998)
29. McLachlan, A.: A mathematical procedure for superimposing atomic coordinates of proteins. Acta Crystallographica Section A 28(6), 656–657 (1972)
30. Medvedev, N.N., Voloshin, V.P., Luchnikov, V.A., Gavrilova, M.L.: An algorithm for three-dimensional Voronoi S-network. Journal of Computational Chemistry 27(14), 1676–1692 (2006)
31. Mendez, R., Leplae, R., Maria, L.D., Wodak, S.J.: Assessment of blind predictions of protein-protein interactions: Current status of docking methods. Proteins: Structure, Function, and Bioinformatics 52, 51–67 (2003)
32. Morris, G.M., Goodsell, D.S., Halliday, R.S., Huey, R., Hart, W.E., Belew, R.K., Olson, A.J.: Automated docking using a Lamarckian genetic algorithm and an empirical binding free energy function. Journal of Computational Chemistry 19(14), 1622–1639 (1998)
33. Moustakas, D.T., Lang, P.T., Pegg, S., Pettersen, E., Kuntz, I.D., Brooijmans, N., Rizzo, R.C.: Development and validation of a modular, extensible docking program: DOCK 5. Journal of Computer-Aided Molecular Design 20, 601–619 (2006)
34. Pei, J., Wang, Q., Liu, Z., Li, Q., Yang, K., Lai, L.: PSI-DOCK: Towards highly efficient and accurate flexible ligand docking. Proteins: Structure, Function, and Bioinformatics 62, 934–946 (2006)
35. Peters, K.P., Fauck, J., Frömmel, C.: The automatic search for ligand binding sites in protein of known three dimensional structure using only geometric criteria. Journal of Molecular Biology 256, 201–213 (1996)

36. Ryu, J., Park, R., Kim, D.-S.: Molecular surfaces on proteins via beta shapes. Computer-Aided Design 39(12), 1042–1057 (2007)
37. Seo, J., Cho, Y., Kim, D., Kim, D.-S.: An efficient algorithm for three-dimensional β-complex and β-shape via a quasi-triangulation. In: Proceedings of the ACM Symposium on Solid and Physical Modeling, June 2007, pp. 323–328 (2007)
38. Shoichet, B.K., Kuntzt, I.D.: Protein docking and complementarity. Journal of Molecular Biology 221, 327–346 (1991)
39. Smith, G.R., Sternberg, M.J.: Prediction of protein-protein interactions by docking methods. Current Opinion in Structural Biology 12, 28–35 (2002)
40. Sousa, S.F., Fernandes, P.A., Ramos, M.J.: Protein-ligand docking: Current status and future challenges. Proteins: Structure, Function, and Bioinformatics 65, 15–26 (2006)
41. Voorintholt, R., Kosters, M.T., Vegter, G., Vriend, G., Hol, W.G.: A very fast program for visualizing protein surfaces, channels and cavities. Journal of Molecular Graphics 7(4), 243–245 (1989)

Kinetic Line Voronoi Operations and Their Reversibility

Darka Mioc[1], François Anton[2], Christopher Gold[3], and Bernard Moulin[4]

[1] National Space Institute, Technical University of Denmark
mioc@space.dtu.dk
[2] Department of Informatics and Mathematical Modelling,
Technical University of Denmark, Richard Petersens Plads,
2800 Kgs. Lyngby, Denmark
fa@imm.dtu.dk
[3] Faculty of Advanced Technologies, University of Glamorgan, Pontypridd,
CF37 1DL, Wales, UK
ChristopherGold@Voronoi.Com
[4] Département d'Informatique, Université Laval, Pavillon Pouliot, Ste Foy,
G1K 7P4, Québec, Canada
moulin@ift.ulaval.ca

Abstract. In Geographic Information Systems the reversibility of map update operations has not been explored yet. In this paper we are using the Voronoi based Quad-edge data structure to define reversible map update operations. The reversibility of the map operations has been formalised at the lowest level, as the basic algorithms for addition, deletion and moving of spatial objects. Having developed reversible map operations on the lowest level, we were able to maintain reversibility of the map updates at higher levels as well. The reversibility in GIS can be used for efficient implementation of rollback mechanisms and dynamic map visualisations. In order to use the reversibility within the kinetic Voronoi diagram of points and open oriented line segments, we need to assure that reversing the map commands will produce exactly the changes in the map equivalent to the previous map states. To prove that reversing the map update operations produces the exact reverse changes, we show an isomorphism between the set of complex operations on the kinetic Voronoi diagram of points and open oriented line segments and the sets of numbers of new / deleted Voronoi regions induced by these operations, and its explanation using the finite field of residual classes of integers modulo 5: $F_5 = \mathbb{Z}/5\mathbb{Z}$. We show also an isomorphism between the set of complex operations on the kinetic Voronoi diagram of points and open oriented line segments and the set of differences of new and deleted Quad-Edge edges induced by these operations, and its explanation using the commutative ring $\mathbb{Z}_{15} = \mathbb{Z}/15\mathbb{Z}$. We show finally the application of these theoretical results to the logging of a kinetic line Voronoi data structure.

M.L. Gavrilova et al. (Eds.): Trans. on Comput. Sci. IX, LNCS 6290, pp. 139–165, 2010.

1   Introduction

In GIS we would like to be able to undo the operations applied to spatial data. The undo/redo functionality is not easy to implement in GIS software. The two possible approaches are based on either preserving the states of past operations and undoing them or using the reversibility to undo the map update operations. Reversible computing, in a general sense, means computing using reversible operations, that is, operations that can be easily and exactly reversed, or undone. Furthermore, when reversibility is maintained at the highest levels, in computer architectures, programming languages, and algorithms, it provides opportunities for interesting applications such as bi-directional debuggers, rollback mechanisms for speculative executions in parallel and distributed systems, and error and intrusion detection techniques. The reversibility can be maintained at the lowest level, in the physical mechanisms of operation of bit-devices and it could be maintained at higher levels as well. Finally, the two types of reversibility (low-level and high-level) are deeply connected, because, as it turns out, achieving the maximum possible computational performance generally requires explicit reversibility not only at the lowest level, but at all levels of computing devices, circuits, architectures, languages, and algorithms (see [18]). In GIS, the reversibility has not been explored sufficiently yet. The reversibility in GIS can be used for efficient implementation of rollback mechanisms and dynamic animations needed in spatial analysis [14]. While the preservation of the geometry of the past map operations may take a lot of computer space, reversibility might offer a less costly approach for handling the cancellation of GIS operations.

In order to use the reversibility for undoing the map updates within the kinetic Line Voronoi diagrams context (which is different from the normal practice in commercial GISs), we need to assure that reversing the map commands will produce exactly the reverse changes in the map equivalent to the previous map states.

We can do it by showing the isomorphism between the set of operations on kinetic Line Voronoi diagrams and the finite field of residual classes of integers modulo 5: $F_5 = \mathbb{Z}/5\mathbb{Z}$.

Even though the finite field of residual classes of integers modulo a prime number: $F_p = \mathbb{Z}/p\mathbb{Z}$ has been extensively studied; only a few works deal with its relationships with Georgii Voronoi's work [3,12,20,11,2]. One of these works deals with Voronoi spatial tessellations as a tool for analytical number theory [3]. In this paper, we show an isomorphism between the set of complex operations on the kinetic Voronoi diagram of points and open oriented line segments and the sets of numbers of new / inactivated Voronoi regions induced by these operations, and its explanation using the finite field of residual classes of integers modulo 5: $F_5 = \mathbb{Z}/5\mathbb{Z}$.

We show also an isomorphism between the set of complex operations on the kinetic Voronoi diagram of points and open oriented line segments and the set of differences of new and inactivated Quad-Edge edges induced by these operations, and its explanation using the commutative ring $\mathbb{Z}_{15} = \mathbb{Z}/15\mathbb{Z}$.

We show that the isomorphism between the set of numbers of added /removed Voronoi regions / Quad-Edge edges and the exact same finite field $F_5 = \mathbb{Z}/5\mathbb{Z}$.

This proves the isomorphism between the set of operations on kinetic Line Voronoi diagrams and the set of the traces of these operations in terms of Voronoi regions / Quad-Edge edges. We show finally, the application of these theoretical results to the logging of a kinetic line Voronoi data structure.

2   Preliminaries: The Voronoi Diagram of Points and Open Oriented Segments

The Voronoi diagram for a set of map objects (points and straight line segments) is the tessellation of space where each map object is assigned an influence zone (or Voronoi region), that is the set of points closer to that object than to any other object (see [8] and Figure 1).

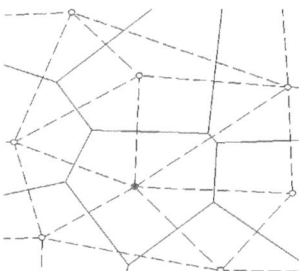

Fig. 1. A Voronoi diagram

More formally, let us first introduce the definition of the Voronoi diagram for a set of sites (i.e. objects or subsets) in the Euclidean affine plane. Let us now consider a set of points and open oriented straight line segments $\mathcal{O}=\{O_1, ..., O_s\}$ in the Euclidean plane, such that the extremities of each one of the open oriented straight line segments belong also to $\mathcal{O}$. We recall that an open oriented straight line segment is the open set formed from a (close straight line) segment $[AB]$ by removing its extremities A and B and considering as positive orientation along the line (AB) the orientation from A to B. The distance from a point M to an object O_i is defined as: either the Euclidean distance between the two points if O_i is a point, or ∞ if the point M lies on the right of O_i, or $d(M, O_i) := \inf_{P \in O_i} d_e(M, P)$ where d_e denotes the Euclidean distance between two points, otherwise. The Voronoi diagram for a set of points and open oriented (straight) line segments is a generalized Voronoi diagram. Let's now introduce the definition of a generalized Voronoi diagram (see [16]), in order to be able to introduce the definition of the Voronoi diagram for a set of points and oriented line segments as a generalized Voronoi diagram. Let S be the metric space in which we place ourselves (typically $\mathbb{R}^2$).

Definition 1. *A mapping* $\delta : S \times \mathcal{O} \to \{0,1\}$ *defined by* $(p, O_i) \mapsto \delta(p, O_i)$ *such that:*

$$\delta(p, O_i) = \begin{cases} 1, & \text{if } p \text{ is assigned to } O_i \\ 0, & \text{otherwise} \end{cases}$$

is called an assignment rule.

Under an assignment rule δ, we consider the set $v(O_i)$ of points assigned to O_i, and the set $e(O_i, O_j)$ of points assigned to both O_i and O_j with $i \neq j$.

Definition 2. *A Voronoi tessellation is a set* $V(\mathcal{O}, \delta, S)$ *such that the assignment rule* δ *satisfies the following two conditions:*

- *every point in* S *is assigned to at least one element of* $\mathcal{O}$ *i.e.,* $\forall p \in S, \sum_{i=1}^{s} \delta(p, O_i) \geq 1$;
- *the set* $e(O_i, O_j)$ *pertains to the boundaries of* $v(O_i)$ *and of* $v(O_i)$, *i.e.,* $\forall \varepsilon \in \mathbb{R}^+, \forall p \in e(O_i, O_j)$:
  - $N_\varepsilon(p) \cap [v(O_i) \setminus e(O_i, O_j)] \neq \emptyset$ *and*
  - $N_\varepsilon(p) \cap [S \setminus v(O_i)] \neq \emptyset$ *and*
  - $N_\varepsilon(p) \cap [v(O_j) \setminus e(O_i, O_j)] \neq \emptyset$ *and*
  - $N_\varepsilon(p) \cap [S \setminus v(O_j)] \neq \emptyset$.

Indeed, the first condition implies that the elements in $V(\mathcal{O}, \delta, S)$ are collectively exhaustive i.e., $\bigcup_{i=1}^{n} v(O_i) = S$.

The definitions of $v(O_i)$ and of $e(O_i, O_j)$ together with the second condition imply that the elements in $V(\mathcal{O}, \delta, S)$ are mutually exclusive except for boundaries i.e.,

$[v(O_i) \cap v(O_j)] \setminus e(O_i, O_j) = \emptyset$ for all $i \neq j$.

Definition 3. *The Voronoi region of* O_i *is:*
$v(O_i) = \{p \in S \mid \delta(p, O_i) = 1\}$.

Definition 4. *The Voronoi edge of* O_i *and* O_j *with* $i \neq j$ *is:*

$$e(O_i, O_j) = \{p \in S \mid \delta(p, O_i) = \delta(p, O_j) = 1\}.$$

Definition 5. *The Voronoi tessellation is the set*
$V(\mathcal{O}, \delta, S) = \{V(O_1), ..., V(O_n)\}$.

We designate this tessellation the generalized Voronoi diagram generated by the generator set $\mathcal{O}$ with assignment rule δ in space S, and $V(O_i)$ the generalized Voronoi region associated with O_i. We call the assignment rule δ that generates a generalized Voronoi diagram, the Voronoi generation assignment rule, or shortly the $V-$assignment rule.

Definition 6. *The Voronoi diagram for a set of points and open oriented straight line segments in the Euclidean plane is a generalized Voronoi diagram where the space is the Euclidean plane, the generator set is comprised of points and/or* *pairs of open oriented straight line segments in the Euclidean plane such that their extremities belong also to the generator set, and the generator assignment rule is as follows.*

If O_i is a point, then
$$\delta(p, O_i) = \begin{cases} 1, & \text{if } d(p, O_i) \leq d(p, O_j), \forall j \\ 0, & \text{otherwise} \end{cases}.$$
If O_i is an open oriented line segment, then
$$\delta(p, O_i) = \begin{cases} 1, & \text{if } d(p, O_i) \leq d(p, O_j), \forall j \text{ and} \\ & p \text{ is on the left of or on } O_i \\ 0, & \text{otherwise} \end{cases}.$$

The Voronoi cell $V(O_i)$ of O_i is the set of points that are closer to or at the same distance with respect to (in the sense of the distance between a point and an object defined just above) O_i than to other sites $O_j : j \in [1, n] \setminus \{i\}$. Then let us introduce the definition of the Delaunay graph of a set of sites (or objects) in the Euclidean affine plane.

Definition 7. *The Delaunay triangulation of $\mathcal{O}$ is the geometric dual of the Voronoi diagram of $\mathcal{O}$: two sites of $\mathcal{O}$ are linked by an edge in the Delaunay triangulation if and only if their cells are incident in the Voronoi diagram of $\mathcal{O}$.*

3   Quad-Edge Based Voronoi Data Structure

The algorithm used to construct the Voronoi vertices has been described in [1]. The boundaries between the regions of this tessellation form a net (the Voronoi diagram), whose dual graph (the Delaunay triangulation) stores the spatial adjacency (topology) relationships among objects. Within such a dynamic Voronoi spatial data structure, as developed by Gold [6], the map objects (points and/or line segments) are stored as nodes of the dual spatial adjacency (topology) graph: the Delaunay triangulation. The underlying data structure used is the Quad-Edge data structure [10] (see Figure 2).

Guibas and Stolfi [10] developed a convenient mathematical structure for representing the topological relationships among edges of a pair of dual subdivisions on a two-dimensional manifold[1]. A subdivision of a manifold M is [10] a partition S of M into three finite collections of disjoint parts, the vertices (denoted by $\mathcal{VS}$), the edges (denoted by $\mathcal{ES}$) and the faces (denoted by $\mathcal{FS}$) with the following properties:

- Every vertex is a point of M,
- Every edge is a line of M,
- Every face is a disk of M, and
- The boundary of every face is a closed path of edges and vertices.

---

[1] A two-dimensional manifold is a topological space with the property that every point has an open neighbourhood which is a disk.

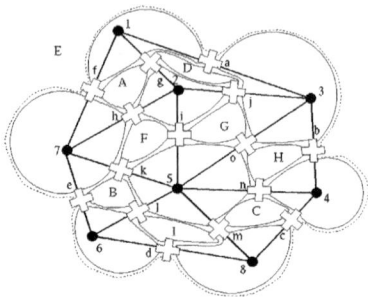

Fig. 2. The Quad-Edge data structure for the Voronoi diagram of Figure 1

The Quad-Edge data structure was used for computing the line Voronoi diagram [8], which is the basis of the dynamic Voronoi data structure for points and line segments. The Quad-Edge data structure is the implementation of an edge algebra [10], which is the mathematical structure that defines the topology of any pair of dual subdivisions on a two-dimensional manifold. In the context of the application of the Quad-Edge data structure to the computation of Voronoi diagrams, both a primal planar graph (the Voronoi diagram) and its dual graph (the Delaunay triangulation) are stored in the Quad-Edge data structure - see [10].

A directed edge of a subdivision P is an edge of P together with a direction along it (see page 80 in [10]). Since directions and orientations can be chosen independently, for every edge of a subdivision there are four directed, oriented edges [10]. For any oriented directed edge e we can define unambiguously its vertex of *origin e.Org*, its *destination, e.Dest*, its *left face*, and its *right face*. The flipped version *e.Flip* of an edge e is the same unoriented edge taken with *opposite orientation* and same direction. The edge *e.Rot* is the edge of the dual subdivision that goes from the right face of e to the left face of e and oriented so that moving counterclockwise around the right face of e corresponds to moving counterclockwise around the origin of *e.Rot*. The *next edge with the same origin, e.Onext* is defined as the one immediately following e (counterclockwise) in the ring of edges out of the origin of e (see Figure 3). Edge functions (see Figure 3) allow the traversal of the pair of dual subdivisions.

3.1   Edge Algebra

An Edge algebra is the mathematical structure used for representing simultaneously a pair of dual subdivisions [10] (in our use of the Quad-Edge data structure, the Delaunay triangulation and the Voronoi diagram). It captures all the topological properties of a subdivision [10]. The topology of the subdivision is completely determined by its edge algebra, and vice versa.

The main advantage of the Quad-Edge data structure is that all the construction and modification of planar graphs can be done using two basic topological operators, and the complex topological operations built from these two basic topological operators:

- e := MakeEdge[] creates an edge e to a newly created data structure representing an empty manifold; and
- Splice[a,b] joins or separates the two edge rings a Org (denoted u) and b Org (denoted v), and independently, the two dual edge rings a Left (denoted α) and b Left (denoted β, see Figure 5).

The *symmetric* of e, e.Sym corresponds to the same undirected edge with the *opposite direction* but the same orientation as e.

Edge functions (see Figure 3) allow the traversal of the pair of dual subdivisions (see Figure 3). The *next counterclockwise edge with the same left face*, denoted by e.Lnext, is defined as the first edge we encounter after e when moving along the boundary of the face $F = e.Left$ in the counterclockwise sense as determined by the orientation of F.

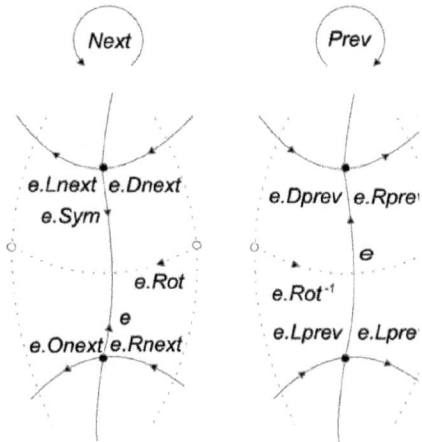

Fig. 3. The edge functions (adapted from Guibas and Stolfi [10])

As shown in the top part of Figure 4, each branch of the Quad-Edge is part of a loop around a Delaunay vertex/Voronoi face, or around a Delaunay triangle/Voronoi vertex. The lower part of Figure 4 shows the corresponding Delaunay/Voronoi structure, where (a,b,c) are Quad-Edges, and (1,2,3) are Delaunay vertices.

3.2 Delaunay/Voronoi Quad Edge Equivalence

Two subdivisions S and S^* are said to be the *dual* [10] of each other if for every directed and oriented edge e of either subdivision there is another edge e.Dual (that is defined as the dual of e and isolated by parenthesis in the following expressions) of the other subdivision such that:

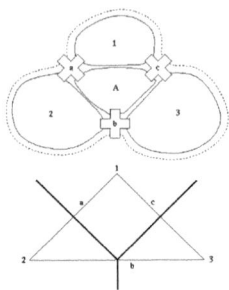

Fig. 4. A simple Voronoi diagram and its corresponding Quad-Edge

- the dual of $e.Dual$ is e: $(e.Dual).Dual = e$,
- the dual of the symmetric of e is the symmetric of $e.Dual$: $e.Sym.Dual = (e.Dual).Sym$,
- the dual of the flipped version of e is the symmetric of the flipped version of $e.Dual$: $e.Flip.Dual = (e.Dual).Flip.Sym$, and
- moving counterclockwise around the left face of e in one subdivision is the same as moving clockwise around the origin of $e.Dual$ in the other subdivision: $e.Lnext.Dual = (e.Dual).Onext^{-1}$.

The dual of an edge e is the edge of the dual subdivision that goes from the (vertex corresponding to the) left face of e to the (vertex corresponding to the) right face of e but taken with orientation opposite to that of e. The definition of the dual of an edge allows to define the operation Rot: the rotated version of an edge e is the dual of e directed from $e.Right$ to $e.Left$ and oriented so that moving counterclockwise around the right face of e corresponds to moving counterclockwise around the origin of $e.Rot$. More concisely, $e.Rot = e.Dual.Flip.Sym = e.Flip.Dual$.

This allows all the edge functions to be expressed using three basic primitives, $Flip$, Rot, and $Onext$ described above [10]. An edge algebra is [10] an abstract algebra $(E, E*, Onext, Flip, Rot)$ where E and E^* are arbitrary finite sets (of edges), and $Onext$, Rot, and $Flip$ are functions on E and E^* satisfying the following properties:

- $e.Rot^4 = e$;
- $e.Rot.Onext.Rot.Onext = e$;
- $e.Rot^2 \neq e$;
- $e \in \mathcal{ES} \Leftrightarrow e.Rot \in \mathcal{ES}^*$;
- $e \in \mathcal{ES} \Leftrightarrow e.Onext \in \mathcal{ES}$;
- $e.Flip^2 = e$;
- $e.Flip.Onext.Flip.Onext = e$;
- $e.Flip.Onext^n \neq e$ for any n;
- $e.Flip.Rot.Flip.Rot = e$; and
- $e \in \mathcal{ES} \Leftrightarrow e.Flip \in \mathcal{ES}$.

The Quad-Edge traversal operations are based on the edge algebra $(E, E*, Onext, Flip, Rot)$, and their expression as composition of the basic primitives [10]. $Onext$, $Flip$, and Rot will be presented in the following table. Equivalent definitions separated by $=$ signs have been presented sometimes for some operations (in the left column). Similarly, equivalent decompositions separated by $=$ signs have been presented sometimes for some operations (in the right column).

Table 1. Quad-Edge traversal operations

Quad-Edge Operation	Decomposition using Edge Algebra
$e.Dual$	$e.Flip.Rot$
$e.Dual^{-1}$	$e.Flip.Rot = e.Rot^3.Flip$
$e.Sym = e.Sym^{-1}$	$e.Rot^2$
$e.Flip^{-1}$	$e.Flip$
$e.Rot^{-1}$	$e.Rot^3$
$e.Onext^{-1} = e.Oprev$	$e.Rot.Onext.Rot = e.Flip.Onext.Flip$
$e.Lnext$	$e.Rot^{-1}.Onext.Rot = e.Rot^3.Onext.Rot$
$e.Rnext$	$e.Rot.Onext.Rot^{-1} = e.Rot.Onext.Rot^3$
$e.Dnext$	$e.Rot^2.Onext.Rot^2$
$e.Lprev = e.Lnext^{-1}$	$e.Onext.Rot^2$
$e.Rprev = e.Rnext^{-1}$	$e.Rot^2.Onext$
$e.Dprev = e.Dnext^{-1}$	$e.Rot^{-1}.Onext.Rot^{-1} = e.Rot^3.Onext.Rot^3$

3.3   Basic Topological Operations in the Quad-Edge Data Structure

The main advantage of the Quad-Edge data structure is that all the construction and modification of planar graphs can be done using two basic topological operators (see Table 2), and the complex topological operations are built from these two basic topological operators. These complex topological operations are presented in the Table 3. The two basic operators modify the graph locally. Locality of the Quad-Edge operations will be studied in detail in the next subsection.

3.4   Locality of the Quad-Edge Data Structure

In vector based GIS systems, the maintenance of topology is performed through batch operations, that are global, i.e. by altering all the objects in the map. In contrast, within the Voronoi spatial data structure, the topology is maintained locally when objects are added and deleted. Indeed, only the neighbours of the object being added/removed may be altered by the topology maintenance. In this section, we will see why the operations on the Quad-Edge data structure have a local scope. In order to prove the locality of the Quad-Edge data structure,

Table 2. Basic Quad-Edge topological operators

Operation	Description
e := MakeEdge[]	Creates an edge e to a newly created data structure representing an empty manifold
Splice[a,b]	Joins or separates the two edge rings a.Org and b.Org, and independently, the two dual edge rings a.Left and b.Left (see Figure 5)

Table 3. Complex Quad-Edge topological operators

Operation	Description
e := Connect[a,b]	Adds a new edge e connecting the destination of a to the origin of b, in such a way that a.Left = e.Left = b.Left
DeleteEdge[e]	Disconnects the edge e from the rest of the data structure
Swap[e]	Rectifies e in order to respect the empty circumcircle criterion

we need to prove the locality of its topological operations. There is only one topological operation within the Quad-Edge data structure: the Splice operation. In the next paragraph, we study the scope of the Splice operation.

$Splice[a, b]$ constructs a new edge algebra $A' = (E, E^*, Onext', Rot, Flip)$ from an existing edge algebra $A = (E, E^*, Onext, Rot, Flip)$. The only difference between A and A' is their $Onext$ edge function. $Onext'$ differs from $Onext$ in the following ways:

– interchange the values of the next edge with the same origin of a with the next edge with the same origin of b:
  • the edge immediately following a with the same origin in A' is the edge immediately following b with the same origin in A (see Figure 5): $a.Onext' = b.Onext$, and
  • the edge immediately following b with the same origin in A' is the edge immediately following a with the same origin in A (see Figure 5): $b.Onext' = a.Onext$;

– interchange the values of the next edge with same origin of $\alpha = a.Onext.Rot$ (see Figure 5) with the next edge with same origin of $\beta = b.Onext.Rot$ (see Figure 5):

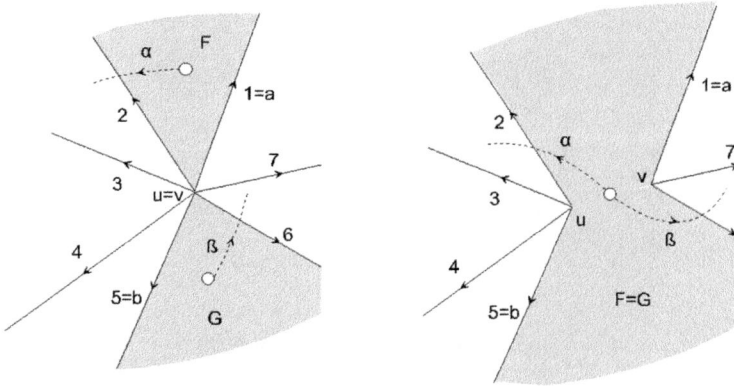

Fig. 5. The Splice topological operator

- the edge immediately following α with same origin in A' is the edge immediately following β with same origin in A: $\alpha.Onext' = \beta.Onext$, and
- the edge immediately following β with the same origin in A' is the edge immediately following α with the same origin in A: $\beta.Onext' = \alpha.Onext$; and

– for each change of the value of the next edge with the same origin of some edge e (i.e. $e.Onext' = f$), redefine the next edge with same origin of the flipped version of f ($f.Flip.Onext'$) to be the flipped version of e:

- the edge immediately following $b.Onext.Flip$ with same origin in A' is the flipped version of a in A: $(b.Onext.Flip).Onext' = a.Flip$,
- the edge immediately following $a.Onext.Flip$ with the same origin in A' is the flipped version of b in A: $(a.Onext.Flip).Onext' = b.Flip$,
- the edge immediately following $\beta.Onext.Flip$ with the same origin in A' is the flipped version of α in A: $(\beta.Onext.Flip).Onext' = \alpha.Flip$, and
- the edge immediately following $\alpha.Onext.Flip$ with same origin in A' is the flipped version of β in A: $(\alpha.Onext.Flip).Onext' = \beta.Flip$.

Now, we can conclude that the scope of the Splice operation is limited to the edges a, b, α, and β, and the flipped version of the next edges with the same origin of a, b, α, and β ($a.Onext.Flip$, $b.Onext.Flip$, $\alpha.Onext.Flip$, $\beta.Onext.Flip$). We conclude that the Splice operation has a local scope, and therefore, that the Quad-Edge data structure has a local scope.

This property of the Quad-edge data structure imposes the following definition of edge modifications due to the operations on the data structure:

– newly created Quad-edges, when a new point is inserted into the structure (one Voronoi region is created for the newly inserted point, and the neighbouring regions are modified);

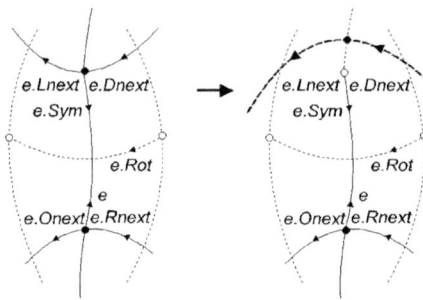

Fig. 6. A modified Quad-edge

- deleted Quad-edges, where the edges belonging to the deleted point are deleted (the deletion of the point or line segment in the Voronoi diagram removes its belonging Voronoi cell, and consequently the deletion and the modification of the edges occur); and
- modified Quad-edges, under stolen area interpolation (see [9]), and triangle switches. Modified edges are edges with the same ID as before, only one or two vertices are changed (see Figure 6).

3.5   Reversibility of the Quad-Edge Operations

Within vector based GIS systems, the operations of maintenance of topology are not reversible. The topology of the entire map is computed by batch operations. The only way to revert to a previous state of the entire map is to store the map before and after each set of batch operation. There is no possibility to revert to a previous local state (i.e. to reverse the topology operation on a region of a map). In this section, we will see that the set of operations on the Quad-Edge data structure is equal to its closure under inversion. This is what we mean by reversibility of the Quad-Edge operations. From the reversibility of the operations on the Quad-Edge data structure, we will prove in a later section that the set of the operations on the Voronoi data structure is also closed by inversion. In order to prove the closure of the set of operations on the Quad-Edge data structure, we need to prove that the inverse of each one of the operations on the Quad-Edge data structure pertains to the set of operations on the Quad-Edge data structure. Before doing this, we prove the reversibility of the operations on the edge algebra, on which the Quad-Edge data structure is based.

In order to prove the reversibility of the operations on an edge algebra $A = (E, E^*, Onext, Rot, Flip)$, we need to prove the reversibility of the primitive edge functions $Onext$, Rot, and $Flip$.

The $Onext$ edge function maps an edge of the primal E to an edge of the primal E, or an edge of the dual E^* to an edge of the dual E^*:

Onext:

$E \rightarrow E$

$E^* \rightarrow E^*$

$e \mapsto e.Onext.$

In both cases, the image of e is the edge immediately following e with same origin.

The reverse of *Onext* is also an edge function: it is *Oprev*, and its decomposition using edge algebra primitive edge functions is *Rot.Onext.Rot*:

$Onext^{-1} = Rot.Onext.Rot$:

$E \rightarrow E$

$E^* \rightarrow E^*$

$e \mapsto e.Rot.Onext.Rot = e.Onext^{-1}.$

The *Flip* edge function maps an edge of the primal E to an edge of the primal E, or an edge of the dual E^* to an edge of the dual E^*:

$Flip$:

$E \rightarrow E$

$E^* \rightarrow E^*$

$e \mapsto e.Flip.$

In both cases, the image of e is the flipped version of e (i.e. the edge connecting the same vertices as e, with the same direction as e, but with opposite orientation).

The reverse of *Flip* is also an edge function: it is *Flip* itself: *Flip* is an involution ($Flip^2 = id$):

$Flip^{-1} = Flip$

$E \rightarrow E$

$E^* \rightarrow E^*$

$e \mapsto e.Flip^{-1} = e.Flip.$

The *Rot* edge function maps an edge of the primal E to an edge of the dual E^*, or an edge of the dual E^* to an edge of the primal E^*:

Rot:

$E \rightarrow E^*$

$E^* \rightarrow E$

$e \mapsto e.Rot.$

The reverse of *Rot* is also an edge function: it is Rot^3, and its decomposition using edge algebra primitive edge functions is *Rot.Rot.Rot*:

$Rot^{-1} = Rot^3$:

$E^* \rightarrow E$

$E \rightarrow E^*$

$e \mapsto e.Rot^3 = eRot.Rot.Rot.$

The reversibility of the other edge functions results from application of the reversion of the composition of applications. Let f and g denote two edge functions, then $(g \circ f)^{-1} = f^{-1} \circ g^{-1}$. Let us apply it to some edge function: the reverse of the *Lprev* edge function whose decomposition into primitive edge functions is $Onext.Rot^2$, is $Lprev^{-1} = (Onext.Rot^2)^{-1} = (Rot^2)^{-1}.(Onext)^{-1} = Rot^2.Rot.Onext.Rot = Rot^3.Onext.Rot$, which is an edge function that can be also written as composition of the primitive edge functions $Onext$, Rot, and $Flip$. The same reasoning can be applied to any edge functions in order to prove its reversibility.

Let us examine now the reversibility of the quad-edge topological operations. The quad-edge topological operations are:

e:=MakeEdge[] creates a new data structure representing a subdivision of the sphere, where apart from orientation and direction, e is the only edge of the subdivision, and e is not a loop [10]. Its reverse would be DeleteEdge[e], deleting a data structure representing a subdivision of the sphere, where apart from orientation and direction, e is the only edge of the subdivision, and e is not a loop. Therefore, the edge e must have been disconnected from all the edges connected to it before calling the DeleteEdge operation.

Splice[a,b] is self reversible: $(Splice[a, b])^{-1} = (Splice[a, b])$. These operations can be written in the terms of edge algebra[2].

In the next section the further formalization of the operations within the Voronoi diagram will be presented.

4   The Operations on the Dynamic Voronoi Data Structure

The complex operations [6] can be decomposed into sequences of atomic actions. Each atomic action in a complex operation executes the geometric algorithm for addition, deletion or change of objects and corresponding Voronoi cells.

The *atomic actions* are:

- the *Split* action inserts a new point into the structure by splitting the point nearest to the location in two (see Figure 7);

- the *Merge* action deletes the selected point by merging it with its nearest neighbour (see Figure 7);
- the *Switch* action is performed when a point moves and a topological event occurs (i.e. the moving point enters or exits a circle circumscribed to a Delaunay triangle, switching[3] the common boundary of two adjacent triangles (see Figure 8);

---

[2] In section 3.1 of [10], it is shown that the topology of a subdivision is completely determined by its edge algebra.

[3] The *Switch* action will be used in the construction of the *Move* action. The *Move* (topological event) action moves the selected point from its current position to a new position or until the next topological event;

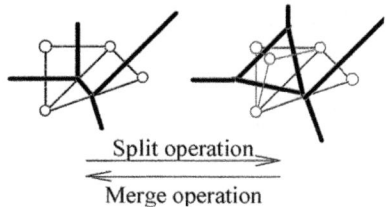

Fig. 7. The topological changes due to the Split and Merge operations

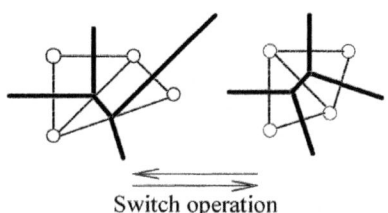

Switch operation

Fig. 8. The topological changes due to the Switch operation

- the *Link* action adds a line segment[4] between the points obtained after a Split (see Figure 9). A Link must occur after a Split, and adds a line segment between the point selected for splitting and the newly created point; and

- the *Unlink* action removes the selected line segment (see Figure 9). The Unlink must occur before the Merge, and removes the line segment between the selected point and its nearest object.

These actions compose the set of atomic actions of the dynamic spatial Voronoi data structure [13].

4.1 Topological Changes in the Voronoi Data Structure Induced by the Atomic Actions

The map changes produced by the atomic actions on this data structure are the changes in the spatial adjacency relationships among spatial objects. Therefore, the only map state changes of the dynamic Voronoi data structure produced by events are the changes in the Delaunay triangulation/Voronoi diagram preserved in the Quad-Edge structure. These events are ruled by the Delaunay triangulation empty circumcircle criterion.

---

[4] A line is composed of two open oriented segments, whose Voronoi regions are on each side of the line segment, having the line segment as a common boundary (see Figure 9).

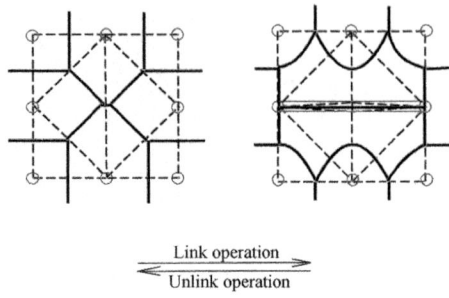

Link operation

Unlink operation

Fig. 9. The topological changes due to the Link and Unlink operations

When a point comes inside a circumcircle or exits from a circumcircle - a "topological event" occurs - the boundary between the two triangles inscribed in the circumcircle "switches" [19].

Within the dynamic Voronoi spatio-temporal data model, all the operations are local and "kinetic": the addition of a new point is performed by splitting the nearest point into two and moving the newly created point to its destination; and the deletion of an existing point is performed by moving it to its nearest point and merging them. It is easy to see that the two actions described previously are mutually reversible: the reverse of a split being a merge and the reverse of a merge being a split (see Gold [6]). A Split action takes the Voronoi cell of a "parent" point and splits it into two, generating a "child" point that may then be moved to the desired destination.

A Merge action reverses this process, combining two adjacent cells into one.

Each *atomic action* produces different changes in spatial topology. The possible changes are:

– the triangle switches (topological events) changing the corresponding Voronoi edges (see Figure 8),
– the creation of a new map object (point or line) and the corresponding appearance of its Voronoi region, or
– the inactivation of a map object and the corresponding disappearance of its Voronoi region. Objects and spatial adjacency relationships are not removed, but inactivated, in order to be able to record all the history information.

The atomic actions of the dynamic spatio-temporal Voronoi data structure and their reverse actions have been described above (see Figures 7, 8, and 9). We introduce the symbols for each atomic action (N for switch, S for split, M for merge, L for link, and U for unlink) that will be used later for specifications of complex map operations. The switch is self-reversible (its reverse is itself). The split and merge are the reverse of each other. The link and the unlink are the reverse of each other.

Table 4. Split operation

Atomic operation	Quad-Edge implementation

Split (X)
X is the location of the point produced after the split
	e:=Locate[X];
	base:=MakeEdge[];
	base.Dest:=X;
	Splice[base,e];
	base:=Connect[e,base.Sym];
	e:=base.Oprev;
	base:=Connect[e,base.Sym]

Table 5. Merge operation

Atomic operation	Quad-Edge implementation
Merge (e) e is the shortest Delaunay edge connecting the point X to be merged with its nearest neighbour	DeleteEdge[e.Onext.Onext]; DeleteEdge[e.Onext]; DeleteEdge[e]

The topological changes for each atomic action in the map are represented by the numbers of newly created and inactivated spatial adjacency links (i.e. Quad-Edges). Each atomic action is uniquely characterized by the numbers of new Quad-Edge (or Voronoi) edges and inactivated Quad-Edge (or Voronoi) edges. This means that from changes in topology we can determine which atomic action was applied, and vice versa. In other words, the actions on the data structure have a deterministic behaviour.

4.2   The Quad-Edge Implementation of the Atomic Actions

The Quad-edge implementations of the atomic actions *Split* and *Merge* in the Voronoi spatial data structure are given in Tables 4 and 5, where the "Locate (X)" function returns a Delaunay edge that is the boundary of a Delaunay polygon that encloses the point X.

The Quad-edge implementation of the atomic action *Switch* is shown in Table 6. On Figure 10 we can see the topological event caused by "swap" atomic operation.

The Quad-edge implementations of the atomic actions *Link* and *Unlink* in the Voronoi spatial data structure are given in Tables 7 and 8.

Table 6. Topological event operation

Atomic operation	Quad-Edge implementation
Topological event (e) e is the Delaunay edge to be swapped	Swap[e] where e is the "suspect" edge (see Figure 10). A suspect edge is an edge that is no longer valid, because the Delaunay triangulation does not obey the empty circumcircle criterion

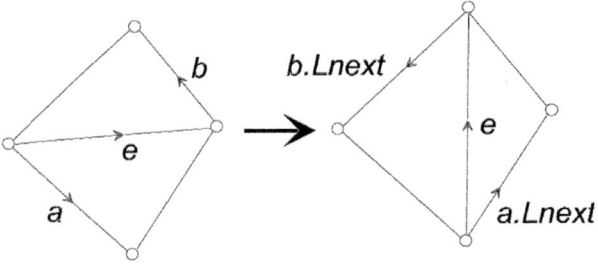

Fig. 10. The topological event caused by "swap" atomic operation

Atomic actions are the basis upon which *map commands* have been built. All the map construction commands [6] of this dynamic Voronoi data structure are complex operations composed of atomic actions (illustrated in Figures 8, 7 and 9). The composition of atomic actions into map commands is provided by syntactic rules. The meaning of the word "syntax" is based on the theory of formal languages and grammars [17]. In the theory of formal languages, the semantics of the basic operations that can be applied on the set of objects is described by a grammar. A grammar provides a set of rules, known as production rules, specifying how the sequence of atomic actions will be applied to the elemental map object (currently a point or a line segment).

Rewriting is a useful technique for defining complex objects by successively replacing parts of elemental map objects using a set of rewrite rules or production rules [17]. Given a set of productions we can generate an infinite number of map objects. In the Voronoi spatial data system there is more than one rule we can apply, and the user is given the freedom of selecting the production rule appropriate to the map update needed. Rewriting context is extended from topological context to include geometrical (spatial) context[5] (position). Therefore rewriting is done sequentially at the specific locations (selected by the user or given by coordinates) and not in parallel (simultaneously) at all possible segments as in

---

[5] In context sensitive systems the selection of the production rule is based on the context of the predecessor. A context sensitive system is needed to model information exchange between neighboring elements [21].

Table 7. Link operation

Atomic operation	Quad-Edge implementation

Link (e) e is the Delaunay edge that will be transformed into two open oriented line segments	a:=e.Onext;
	b:=e.Oprev;
	c:=b.Lnext;
	d:=a.Rprev;
	DeleteEdge[e];
	f:=MakeEdge[];
	f.dest :=(b.Org+c.Dest)/2;
	Splice[b,f];
	e:=MakeEdge[];
	e.org:=f.dest;
	Splice[f,e.Sym];
	g:=Connect[e.Sym,a.Sym];
	i:=Connect[g.Sym,d.Sym];
	h:=Connect[b,f.Sym];
	j:=Connect[c,h.Sym];
	k:=Connect[f,e]

Table 8. Unlink operation

Atomic operation	Quad-Edge implementation

Unlink (e) e is any Delaunay edge that connects the two open oriented line segments to be removed	f:=e.Onext;
	g:=e.Lnext;
	h:=g.Oprev;
	i:=h.Oprev;
	j:=f.Onext;
	k:=e.Onext;
	a:=h.Lnext;
	b:=i.Lnext;
	DeleteEdge[e];
	DeleteEdge[f];
	DeleteEdge[g];
	DeleteEdge[h];
	DeleteEdge[i];
	DeleteEdge[j];
	DeleteEdge[k];
	Connect[a,b]

the L-systems[6]. The implementation of topological and geometrical properties is more suited for spatial representation than for a purely (symbolic) computational representation. For spatial analyses the Voronoi based data structure represents space in a more "intelligent" way[7] (see [5], and [6]).

Therefore, the Voronoi spatial data model as implemented is very similar to the approaches which use developmental systems as a new method of "learning". New models of cellular networks [22] applied to the problems of spatial analyses such as: shortest path, nearest neighbour and navigation showed that the Voronoi data model has a better potential for spatial representation.

4.3   The Map Construction Commands

The update of the Voronoi data structure given by map commands can be interpreted as the execution of production rules which constitute a map grammar [17].

The *map construction commands* are illustrated in Figure 11. On the left side of Figure 11, we can see the map objects on which the map command will be applied, and on the right side, the map objects that have been rewritten. In the graphical illustration of map commands, the topological part of the model is left out for better understanding of the general principle, and also the line-line collisions and their effects are not shown. In Table 14 the reversibility of map commands is shown, and we can see that each map command has its reverse. Furthermore, the map commands can be recognized by the changes between the predecessor (shown in second column of the Table 10) and successor map topology states (shown in third column of the Table 10).

These actions compose the set of atomic actions of the dynamic spatial Voronoi data structure [13].

The *atomic actions* are the basis upon which *complex operations* have been built. All the complex operations [6], [15] of this dynamic Voronoi data structure are complex operations composed of atomic actions. The composition of atomic actions into complex operations is provided by syntactic rules.

The *complex operations* are composed of atomic operations, and the exact decomposition of complex operations into sequences of atomic actions is given in Table 11. The atomic operations are denoted by the symbols (N, S, M, L and U). For example, the operation *"Move a Point"* corresponds to the sequence of movements of the point from its initial position to its destination through all the intersections of its trajectory with circumcircles, and the corresponding triangles

---

[6] The spatial language shows the hierarchical presentation of the production rules and spatial objects. It is similar to interleaved L-systems [21], where one L-system may have other L-systems as its symbols. For example, the move command is a part of any other production rule described as a map command. Even though it has a lot in common with L-systems, we no longer refer to it as L-systems [21].

[7] *"The new theory of emergent behaviour arising in the field of AI considers that intelligence is reflected by the collective behaviours of large numbers of very simple interacting, semi-autonomous individuals, or agents. Whether we take these agents to be neural cells, individual members of a species, or a single person in society, their interactions produce intelligence."* From [21].

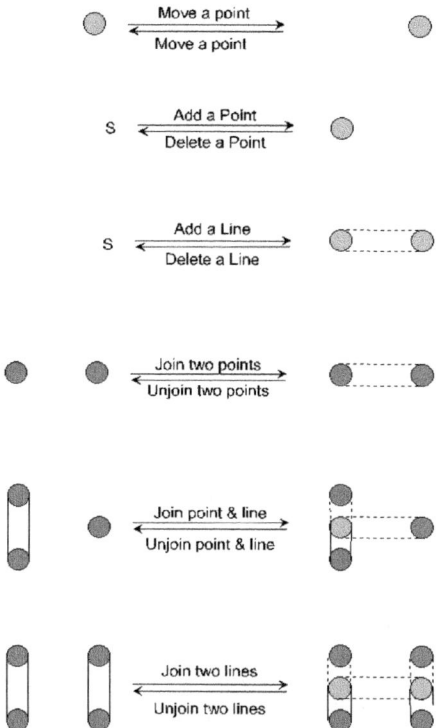

Fig. 11. The map commands (S is the starting symbol in the production rule)

switches (*"N"*) in the Voronoi data structure. The operation *"Move a Point"* is possible in this Voronoi data structure because the Voronoi data structure is kinematic : one point may move at a time, and this point is called the *"moving point"* [7].

In fact, all the operations on this kinematic Voronoi data structure use this concept of the moving point. For example, when a point is to be created at some location, the nearest point from that location is split into two (S term in the decomposition of *"Add a Point"* operation SN^t), and then the newly created point is moved as far as its final destination (N^t term in the decomposition of *"Add a Point"* operation SN^t). In fact, the triangle switch operation incorporates the movement of the moving point to the intersection of the trajectory of the moving point with the circumcircle that induced the triangle switch. In Table 11, the exponents denote how many times the operation is executed repeatedly, e.g. N^t denotes N executed t times, where t denotes the number of topological events. Whenever more than one connected sequence of topological events is executed in a operation, such as in *"Add a Line"* operation (SN^{t_1} SLN^{t_2} ($SLN^{t_{2i+1}}$ $MSLN^{t_{2i+2}}$)), the total number of topological events is broken down into the number of topological events in the first connected sequence (N^{t_1}), the number of topological events in the second connected sequence (N^{t_2}),

Table 9. The reversibility of the map commands

Map construction command	Reverse map construction command
Move a Point	Self-reversible
Add a Point	Delete a Point
Add a Line	Delete a Line
Join 2 Points	Unjoin 2 Points
Join Pt & Line	Unjoin Pt & Line
Join 2 Lines	Unjoin 2 Lines

Table 10. The changes induced by map commands in Voronoi regions

Map construction command	Inactivated Voronoi regions	Newly created Voronoi regions
	c = number of line intersections	
Move a Point	0	0
Add a Point	0	1
Delete a Point	1	0
Add a Line	0	$4 + 5c$
Delete a Line	$4 + 5c$	0
Join 2 Points	0	$2 + 5c$
Unjoin 2 Points	$2 + 5c$	0
Join Point & Line	0	$5 + 5c$
Unjoin Point & Line	$5 + 5c$	0
Join 2 Lines	0	$8 + 5c$
Unjoin 2 Lines	$8 + 5c$	0

and so on. The parameter i denotes the number of times the line segment being added intersects existing line segments. This type of intersection with an existing line segment is called collision. The terms in parentheses are repeated for each intersection with an existing line (i.e. each collision).

We will now briefly explain the decomposition of the other complex operations. The complex operation *"Delete a Point"* is exactly the reverse of *"Add a Point"* operation: the point to be deleted is moved to the location of the nearest point (N^t), and then it is merged with this nearest point (M). The remaining complex operations involve the addition or removal of one or more new line segments. For all these complex operations, the decomposition includes a fixed sequence of atomic actions that is executed only once (the sequence outside the parenthesis), and a sequence that is executed at each collision (replicating sequence). In the case of case*"Add a Line"* and all the join complex operations, the replicating sequence has always the same pattern in terms of atomic operations $((SLN^{t_{2i+1}}MSLN^{t_{2i+2}})$, although the actual indices may vary). This corresponds to the splitting of the existing line $(SLN^{t_{2i+1}})$, the merging of the newly created point (by the S atomic action in this last sequence) with the extremity of the line segment being added (M), and the continuation of the new line segment after collision $(SLN^{t_{2i+2}})$. In the case of *"Delete a Line"* and all the unjoin complex operations, the replicating sequence has always the same pattern in terms of atomic operations $((N^{t_{2i+2}}UMSN^{t_{2i+1}}UM)$, although the actual indices may vary). This is exactly the reverse of the previous replicating sequence (the replication sequence for *"Add a Line"* and all the join complex operations).

Now, we will explain the fixed sequence for all these complex operations. In order to add a line with *"Add a Line"* operation, the nearest point from the starting extremity location has to be split into two (S), then it has to be moved to the starting extremity location (N^{t_1}). Then, the ending extremity has to be created by splitting the starting extremity into two (S). At this point the two extremities must be linked (L) in order to form a line segment. Finally, the ending extremity has to be moved (N^{t_2}) to its expected location. The fixed sequence for *"Delete a Line"* is exactly the reverse of the preceding sequence. In order to join two points with *"Join two points"* operation, the first point must be split into two (S) in order to create the ending extremity of the line segment that starts at the first point. Then, these two points must be linked (L) in order to form a line segment. Then, the ending extremity must be moved (N^{t_1}) to the location of the second point (including eventually the replicating sequence in case of collisions). Finally, the ending extremity must be merged with the second point (M). The fixed sequence for *"Unjoin two points"* operation is exactly the reverse of the fixed sequence for *"Join two points"* operation. The fixed sequences of the remaining complex operations follow immediately from the fixed sequence of *"Join two points"* operation. Indeed, the other join complex operations fixed sequence involve several sequences corresponding to the same atomic actions as the SLN^{t_1} sequence already encountered in the fixed sequence of *"Join two points"* operation. The unjoin complex operations are the exact reverse of their join counterpart.

Table 11. The complex operations and their decomposition into atomic actions

Map construction operation	Decomposition (the terms in parentheses appear at each line-line collision, i = collision index, $i \in \{1, ..., c\}$, c = number of collisions; t, t_x denote numbers of topological events)
Move a Point	N^t
Add a Point	SN^t
Delete a Point	$N^t M$
Add a Line	$SN^{t_1} SLN^{t_2} (SLN^{t_{2i+1}} MSLN^{t_{2i+2}})$
Delete a Line	$(N^{t_{2i+2}} UMSN^{t_{2i+1}} UM) N^{t_2} UMN^{t_1} M$
Join 2 Points	$SLN^{t_1} (SLN^{t_{2i}} MSLN^{t_{2i+1}}) M$
Unjoin 2 Points	$(N^{t_{2i+1}} UMSN^{t_{2i}} UM) N^{t_1} UM$
Join Pt & Line	$SLN^{t_1} (SLN^{t_{2i+1}} MSLN^{t_{2i+2}}) SLN^{t_2} M$
Unjoin Pt & Line	$SN^{t_2} UM (N^{t_{2i+2}} UMSN^{t_{2i+1}} UM) 5N^{t_1} UM$
Join 2 Lines	$SLN^{t_1} SLN^{t_2} (SLN^{t_{2i+2}} MSLN^{t_{2i+3}}) SLN^{t_3} M$
Unjoin 2 Lines	$SN^{t_3} UM (N^{t_{2i+3}} UMSN^{t_{2i+2}} UM) N^{t_2} UMN^{t_1} UM$

Table 12. The discrimination of operations by means of their changes in topology

Op.	New Edges	Old Edges	Difference	Difference set
	t = total number of topological events			
Move P	t	t	0	$\{0\}$
Add P	$t+6$	$t+3$	3	$\{3\}$
Delete P	$t+3$	$t+6$	-3	$\{-3\}$
Add L	$t+23+37c$	$t+11+22c$	$12+15c$	$12_{\mathbb{Z}_{15}} \cap \{z \geq 12\}$
Delete L	$t+11+22c$	$t+23+37c$	$-12-15c$	$3_{\mathbb{Z}_{15}} \cap \{z \leq -12\}$
Join 2P	$t+20+37c$	$t+14+22c$	$6+15c$	$6_{\mathbb{Z}_{15}} \cap \{z \geq 6\}$
Unj. 2P	$t+14+22c$	$t+20+37c$	$-6-15c$	$9_{\mathbb{Z}_{15}} \{z \leq -6\}$
Join P/L	$t+37+37c$	$t+22+22c$	$15+15c$	$0_{\mathbb{Z}_{15}} \cap \{z \geq 15\}$
Unj. P/L	$t+22+22c$	$t+37+37c$	$-15-15c$	$0_{\mathbb{Z}_{15}} \cap \{z \leq -15\}$
Join 2L	$t+54+37c$	$t+30+22c$	$24+15c$	$9_{\mathbb{Z}_{15}} \cap \{z \geq 24\}$
Unj. 2L	$t+30+22c$	$t+54+37c$	$-24-15c$	$6_{\mathbb{Z}_{15}} \cap \{z \leq -24\}$

Table 10 shows the isomorphism between the set of operations on the kinetic Voronoi diagram of points and open oriented line segments and the sets of the numbers of new and deleted Voronoi regions, through subsets of the finite field F_5. Table 12 shows the isomorphism between the set of operations on the kinetic Voronoi diagram of points and open oriented line segments and the sets of the differences between the numbers of new and deleted Quad-Edge (or Voronoi) edges through subsets of the commutative ring $\mathbb{Z}_{15}$. This means that from changes in topology (new/inactivated Quad-Edge edges), we can determine which atomic/complex action was applied, and vice versa, allowing one to log the operations through the differences between the numbers of new/deleted Quad-Edge edges.

Table 13. The reversibility of the atomic actions

Atomic action	Reverse atomic action
Split	Merge
Switch	Switch is self-reversible
Link	Unlink

Table 14. The reversibility of the map commands

Map construction command	Reverse map construction command
Move a Point	Self-reversible
Add a Point	Delete a Point
Add a Line	Delete a Line
Join 2 Points	Unjoin 2 Points
Join Pt & Line	Unjoin Pt & Line
Join 2 Lines	Unjoin 2 Lines

5   Reversibility of the Map Commands in the Dynamic Voronoi Data Structure

For each map command, the reverse map command is composed of reverse atomic actions in exactly the reverse order. Due to the local scope of its spatio-temporal topology, all the atomic actions of the dynamic Voronoi spatio-temporal model are reversible. Indeed, each atomic action has its reverse atomic action shown in the Table 13.

The consequence of the property of reversibility of the atomic actions inside the Voronoi dynamic data structure is that a sequence of atomic actions applied in a map update command can be reconstructed from the predecessor and successor map states. This proves in another way that the atomic actions are reversible: the input can be deduced from the output; or, in other words, computation happens without any loss of information [4].

The resulting complex operations (map commands) are reversible and Table 14, as long as their decomposition into atomic actions is exactly known (including the numbers of topological events and the number of line-line collisions).

6   Conclusions

In this paper we presented formalisation of the reversible operations needed for constructing a Voronoi diagram for points and line segments using the Quad-Edge data structure. These reversible operations are formalized at the lowest level, as the basic algorithms for addition, deletion and moving of spatial objects in the Quad-Edge data structure; defined as the atomic actions. Furthermore, we managed to preserve the reversibility of the map commands that are composed of these atomic actions.

We have shown the isomorphism between the set of operations on the kinetic Voronoi diagram of points and open oriented line segments and the sets of the numbers of new and deleted Voronoi regions, through subsets of the finite field F_5. We have also shown the isomorphism between the set of operations on the kinetic Voronoi diagram of points and open oriented line segments and the sets of the differences between the numbers of new and deleted Quad-Edge (or Voronoi) edges through subsets of the commutative ring $\mathbb{Z}_{15}$. This has been applied to the logging of the operations on the kinetic Voronoi diagram of points and open oriented line segments, which has been the building block of a Voronoi based GIS developped at the Industrial Chair of Geomatics applied to Forestry, Université Laval, Québec, Canada. The same results can be easily generalised to 3D. The applications involve using this isomorphism for the recovery of Voronoi-based spatial databases. The applicability to commercial GISs involves the generation of the Voronoi data structure for all the objects stored in the commercial GIS. The applications of reversible computations in GIS could significantly improve transaction management and rollback functionality.

References

1. Anton, F., Gold, C.M.: An iterative algorithm for the determination of Voronoi vertices in polygonal and non-polygonal domains. In: Proceedings of the 9th Canadian Conference on Computational Geometry (CCCG 1997), Kingston, Canada, pp. 257–262 (1997)
2. Bellman, R.: On a class of functional equations of modular type. Proc. Nat. Acad. Sci. U.S.A. 42, 626–629 (1956)
3. Cui, J.: A decoding algorithm for general $\mathbb{Z}_4$-linear codes. J. Syst. Sci. Complex. 17(1), 16–22 (2004)
4. Frank, M., Knight, T., Margolus, N.: Reversibility in optimally scalable computer architectures. In: The First International Conference on Unconventional Models of Computation, January 1998, pp. 165–182 (1998)
5. Gold, C.M.: Space revisited - back to the basics. In: Proceedings of the Fourth International Symposium on Spatial Data Handling, Zurich, Switzerland, pp. 175–189 (1990)
6. Gold, C.M.: An object-based dynamic spatial data model, and its applications in the development of a user-friendly digitizing system. In: Proceedings of the Fifth International Symposium on Spatial Data Handling, Charleston, pp. 495–504 (1992)
7. Gold, C.M.: Three approaches to automated topology, and how computational geometry helps. In: Proceedings of the Sixth International Seminar on Spatial Data Handling, Edinburgh, Scotland, pp. 145–158 (1994)
8. Gold, C.M., Dakowicz, M.: Kinetic Voronoi/Delaunay Drawing Tools. In: ISVD 2006, pp. 76–84 (2006)
9. Gold, C.M., Remmele, P.R., Roos, T.: Voronoi Diagrams of Line Segments Made Easy. In: Proceedings of the Seventh Canadian Conference in Computational Geometry (CCCG 1995), Québec, Canada, pp. 223–228 (1995)
10. Guibas, L., Stolfi, J.: Primitives for the Manipulation of General Subdivisions and the Computation of Voronoi Diagrams. ACM Transactions on Graphics 4(2), 74–123 (1985)

11. Ireland, F.K., Rosen, I.M.: A classical introduction to modern number theory. Graduate Texts in Mathematics, vol. 84. Springer, New York (1982); revised edition of Elements of number theory
12. Lee, Y., Scheidler, R., Yarrish, C.: Computation of the fundamental units and the regulator of a cyclic cubic function field. Experiment. Math. 12(2), 211–225 (2003)
13. Mioc, D., Anton, F., Gold, C.M., Moulin, B.: Spatio-temporal change representation and map updates in a dynamic Voronoi data structure. In: Proceedings of the Eight International Symposium on Spatial Data Handling, Vancouver, Canada, pp. 441–452 (1998)
14. Mioc, D., Anton, F., Gold, C.M., Moulin, B.: "Time travel" Visualization in a Dynamic Voronoi Data Structure. Cartography and GIS 26(2), 99–108 (1999)
15. Mioc, D., Anton, F., Gold, C.M., Moulin, B.: Map updates in a dynamic Voronoi data structure. In: ISVD 2006, pp. 264–269 (2006)
16. Okabe, A., Boots, B., Sugihara, K., Nok Chiu, S.: Spatial tessellations: concepts and applications of Voronoi diagrams, 2nd edn. John Wiley & Sons Ltd., Chichester (2000); With a foreword by D. G. Kendall
17. Prusinkiewicz, P., Lindenmayer, A.: The Algorithmic Beauty of Plants. Springer, New York (1990)
18. Reversible Computing FAQ,
    http://www.cise.ufl.edu/research/revcomp/faq.html
19. Roos, T.: Dynamic Voronoi diagrams. Ph.D. Thesis, University of Würzburg, Germany (1991)
20. Scheidler, R., Stein, A.: Voronoi's algorithm in purely cubic congruence function fields of unit rank 1. Math. Comp. 69(231), 1245–1266 (2000)
21. Vaario, J.: An Emergent Modeling Method for Artificial Neural Networks. Doctoral dissertation, University of Tokyo, Japan (1993)
22. Vleugels, J., Kok, J.N., Overmars, M.: Motion Planning with Complete Knowledge using a Colored SOM. International Journal of Neural Systems 8(5-6), 613–628 (1997)

High Quality Visual Hull Reconstruction by Delaunay Refinement

Xin Liu and Marina L. Gavrilova

Calgary University, Calgary AB T2N 1N4, Canada
{liuxin,marina}@ucalgary.ca

Abstract. In this paper, we employ Delaunay triangulation techniques to reconstruct high quality visual hulls. From a set of calibrated images, the algorithm first computes a sparse set of initial points with a dandelion model and builds a Delaunay triangulation restricted to the visual hull surface. It then iteratively refines the triangulation by inserting new sampling points, which are the intersections between the visual hull surface and the Voronoi edges dual to the triangulation's facets, until certain criteria are satisfied. The intersections are computed by cutting line segments with the visual hull, which is then converted to the problem of intersecting a line segment with polygonal contours in 2D. A barrel-grid structure is developed to quickly pick out possibly intersecting contour segments and thus accelerate the process of intersecting in 2D. Our algorithm is robust, fast, fully adaptive, and it produces precise and smooth mesh models composed of well-shaped triangles.

Keywords: Visual hull, Delaunay refinement, segment cut, barrel-grid.

1 Introduction

Shape from silhouette (SFS) is a popular image-based modeling technique. It computes a *visual hull*, which is the maximal solid that can reproduce the silhouettes in the input images [15]. The visual hull is a superset of the real shape of the object, which is usually used as an initial model to be further refined by other approaches [13,14,17] or directly used for rendering in virtual reality systems [3,4]. In both applications, the quality of the visual hull model with respect to precision, smoothness, and mesh composition is important. In this paper, we present a novel SFS algorithm for reconstructing high quality visual hulls that is robust, fast and fully adaptive.

SFS has been a hot research topic since it was first proposed by Baumgart [5]. Existing SFS algorithms can be classified coarsely into surface-based approaches and volume-based approaches. Surface-based approaches compute a polygonal surface model of the visual hull as the intersection of all *viewing cones* [12] extruding from the camera centers. Early algorithms [5] directly compute the intersections in 3D space. Matusik *et al.* [22] reduced 3D intersections to simpler intersections in 2D image space. Franco and Boyer [10] reconstructed a polyhedral visual hull in three steps: computing viewing edges, recovering cone intersection edges and triple points, and identifying the faces. Surface-based approaches are fast and precise, but they lack robustness, because intersecting two nearly parallel surfaces is an ill-posed problem, and the mesh model they produce is usually composed of very irregular triangles.

M.L. Gavrilova et al. (Eds.): Trans. on Comput. Sci. IX, LNCS 6290, pp. 166–182, 2010.

Volume-based approaches compute a solid model of the visual hull by *volume carving*. Traditional algorithms [21] use a solid model composed of cubic *voxels*. They reconstruct the visual hull by carving away those voxels projected out of the silhouette regions in one or more images. This process can be accelerated by using an octree structure and performing voxel carving in a coarse-to-fine fashion [25]. The reconstructed voxel model can be converted to a mesh model by a *marching cubes* algorithm [20]. Voxel-based approaches are well known for their robustness. However, they have to face the trade-off between precision and speed, and the mesh model produced by marching cubes is usually composed of over-dense vertices and suffers from discretization artifacts. Boyer and Franco [7] subdivided the volume by irregular tetrahedrons. Liu *et al.* [18] proposed a volumetric algorithm based on a special centripetal pentahedron model.

Boissonnat and Oudot proposed a *Delaunay refinement* paradigm for surface meshing [6]. The main idea is to progressively refine a *Delaunay triangulation* (DT) restricted to the surface by inserting new points which are the intersections between the surface and the Voronoi edges dual to the triangulation's facets. Aganj *et al.* used this idea to compute a four-dimensional *spatio-temporal visual hull* [2]. For static scenes, their algorithm first computes a *volumetric model* and then obtains a mesh model by identifying the *boundary facets* of the volumetric model. It uses *dichotomic searching* to compute the intersections, which is slow and inexact.

In this paper we propose a new Shape from Silhouette (SFS) algorithm. As Aganj *et al.*'s algorithm [2], our algorithm is also based on Boissonnat and Oudot's *Delaunay refinement* paradigm [6], but we do not reconstruct an intermediate volumetric model and we directly compute all exact [1] intersections between a Voronoi edge and the visual hull surface by cutting line segments. Specifically, from a set of calibrated images, our algorithm first computes a sparse set of initial points with a *dandelion model* [19] and builds an initial DT restricted to the visual hull surface. Our algorithm then iteratively refines the triangulation by inserting new sampling points, which are the intersections between the visual hull surface and the Voronoi edges dual to the triangulation's facets, until certain criteria are satisfied. The intersections are computed by *cutting line segments* with the visual hull. With projective transformations, the problem is converted to intersecting an arbitrary line segment with polygonal contours in 2D space. A *barrel-grid* structure is developed which can quickly pick out the possibly intersecting contour segments and thus accelerate the process of intersecting in 2D. In this way, all *intersections* between a Voronoi edge and the visual hull surface can be computed exactly and efficiently. The influence of inserting a point into the DT is local, so the restricted DT can be maintained dynamically by updating only a few *conflict facets*. In the current work, we consider the center-center distance, minimum angle, and circumscribing radius criteria, which work together leading the restricted DT to evolve toward a precise and well-composed mesh model of the visual hull. When the refinement stops, the restricted DT forms a high quality mesh model of the visual hull.

Our algorithm samples the visual hull surface with the Voronoi edges that are perpendicular to the current model's surface, which embodies the best knowledge about

---

[1] Our method is able to compute the exact intersections theoretically, but we used floating-point number computations with limited precision in programming.

the visual hull that has been discovered. In this way, the model vertex computation is generally well-posed, and thus our algorithm is robust. The termination of the algorithm implies all criteria are satisfied, so the reconstructed model is precise and well-composed. The visual hull is defined on polygonal contours (rather than discrete pixels, as in Liang and Wong's work [16]) and all computations are exact, so the output model is smooth. The efficient dynamic DT algorithm and quick intersection algorithm make our algorithm reasonably fast. To summarize, the contributions of this paper are:

- an efficient and robust algorithm for reconstructing high quality visual hulls,
- an algorithm to compute all exact intersections between a Voronoi edge and the visual hull surface by cutting line segments,
- a barrel-grid structure and the related algorithms for fast line segment/polygon intersection in 2D.

2   Preliminaries

To make the paper self-inclusive, we first make a brief review of some basic concepts about 3D Voronoi diagram and Delaunay triangulation.

Let $E = \{P_1, P_2, \cdots, P_n\}$ be a set of points in $\mathbb{R}^3$. The *Voronoi cell* associated to a point P_i is the 3D region that is closer from P_i than from any other points in E:

$$\mathcal{V}(P_i) = \left\{ P \in \mathbb{R}^3 : \forall j \Rightarrow \|P, P_i\| \leq \|P, P_j\| \right\}. \tag{1}$$

The *Voronoi diagram* (VD) of E, $Vor(E)$, is a partition of the space $\mathbb{R}^3$ with the Voronoi cells $\mathcal{V}(P_i)$, $i = 1, 2, \cdots, n$. The *Delaunay triangulation* (DT) of the point set E, $Del(E)$, is a partition of the convex hull of E by tetrahedrons satisfying the *empty circumsphere* rule, i.e., the circumsphere through the vertices of a tetrahedron contains no point of E in its interior. If the points in E are in general positions, i.e., no five points are cospherical, $Del(E)$ is unique and dual to $Vor(E)$: Each vertex, edge, facet, and tetrahedron of $Del(E)$ is dual to a cell, facet, edge, and vertex of $Vor(E)$, respectively. Unlike the 2D version of Delaunay triangulation, the 3D DT does not ensure that the minimum angle in each DT is maximized. However, in most cases the 3D DT has a tendency to favor equilateral tetrahedra, which are preferable for meshing.

The Delaunay triangulation of a point set E restricted to a surface S, $Del_{|S}(E)$, is the sub-complex of $Del(E)$ consisting of the facets of $Del(E)$ whose dual Voronoi edges intersect S [6]. The restricted DT is a polygonal approximation of the surface S. Figure 1 illustrates these concepts in 2D space.

3   Algorithm Framework

To help readers maintain a mental image of the proposed Shape from silhouette (SFS) algorithm, we first describe the algorithm in a high-level in this section. The details of involved techniques are then presented in the following sections.

As shown in the pseudo code below, our SFS algorithm dynamically maintains three data structures during the process of reconstruction: the DT, the restricted DT, and the

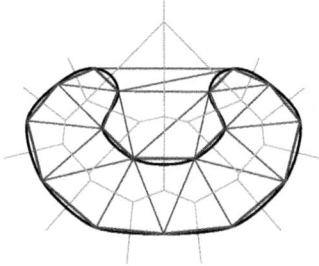

Fig. 1. VD and DT illustrated in 2D. The black curve denotes the object's surface; the VD is shown in blue; the DT is shown in red; the restricted DT is shown by the bold red line segments.

queue of bad facets sorted according to their quality. To make the description clear, we divide the process into four steps. In the *first step*, image processing, the algorithm first extracts polygonal contours and then computes the barrel-grid structures to get ready for quick line segment/contour intersection in 2D, as detailed in Sec. 6. In the *second step*, the algorithm computes an initial DT restricted to the visual hull surface. This is done by three sub-steps. First, a set of initial points E is computed, as is discussed in Sec. 4. Second, DT is applied to the initial points to obtain $Del\,(E)$. Third, we scan the facets of $Del\,(E)$ to initialize the DT restricted to the visual hull surface $Del_{|VHS}\,(E)$ and the bad facet queue. The *third step* refines the triangulation by iteratively inserting new sampling points to the point set E, which are the intersections between the visual hull surface and the Voronoi edges dual to the triangulation's facets, until all criteria are satisfied. When a point is inserted into the DT, its influence is local. Therefore, before insertion, we first identify the *conflict facets* to be influenced, and delete them from $Del\,(E)$, $Del_{|VHS}\,(E)$, and the bad facet queue. Then the new point is inserted into $Del\,(E)$, after which the new facets are checked to restore $Del_{|VHS}\,(E)$ and the bad facet queue. The problem of computing the intersections between the visual hull surface and the Voronoi edges is discussed in Sec. 5; the refinement criteria are discussed in Sec. 7. Finally, in the *fourth step*, a high quality mesh model is output which is directly available from $Del_{|VHS}\,(E)$.

```
SFS()
{
    // Step 1: image processing
    for each image {
        ExtractSilhouetteContour ( );
        ComputeBarrelGrid( );
    }
    // Step 2: Initialize the restricted DT
    E = InitialPoints( );
    DT = DelaunayTriangulation(E);
    RDT = ∅; // the restricted DT
    FacetQ = ∅; // the queue of bad facets
    for each facet f in DT
```

```
    CheckFacet (f);
  // Step 3: refine the restricted DT
  while ( ! Empty (FacetQ) ) {
    f = PopFront (FacetQ);
    VoroEdge = Dual(f);
    NewPoints = IntersectWithVHSurf(VoroEdge);
    for each point P in NewPoints {
      // pre-insertion processing
      ConflictFacets = FindConflictFacets (DT, P);
      remove facets in ConflictFacets from FacetQ, DT and RDT;
      // insertion
      InsertVertex (DT, P);
      // post-insertion processing
      for each new facet f'
        CheckFacet (f');
    }
  }
  // Step 4: output
  output the mesh model from RDT;
}
```

As shown below, the function CheckFacet() first computes the Voronoi edge dual to the facet of the Delaunay triangulation. It then computes the intersections between the Voronoi edge and the visual hull surface. If there are one or more intersections, the facet will be added into the restricted DT. If the facet belongs to the restricted DT and it does not satisfy the criteria, the facet will be inserted into the bad facet queue.

```
CheckFacet (facet f)
{
  VoroEdge = Dual(f);
  if ( IntersectWithVHSurf (f) ≠ ∅) {
    add f to RDT;
    if ( ! CriteriaSatisfied (f))
      Insert (FacetQ, f);
  }
}
```

4   Initial Point Computation

At the beginning of reconstruction, we know nothing about the destination model. In this case we intend the sampling points to be computed under well-posed conditions and to be distributed on the visual hull surfaces as evenly as possible. Therefore we use a *dandelion model* first proposed by Liu et al. [19] to sample the visual hull surface. The process is as follows: First, a bounding box and its center O is computed by the algorithm proposed by Mulayim et al. [23]. Then an n-geodesic sphere centered at O

is constructed and a pencil of *sampling rays* emitted from O and through the vertices of the geodesic sphere are obtained, as shown in Fig. 2. Finally, we intersect the rays with the visual hull surface and all intersections form the initial point set.

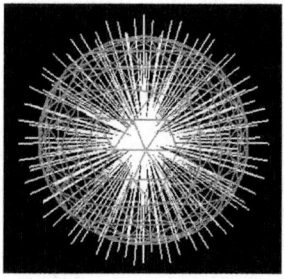

Fig. 2. The dandelion model [19]

A *geodesic sphere* is a polygonal approximation of a sphere. It can be constructed from a regular icosahedron by iteratively subdividing its facets. For each subdivision of the model, each edge is broken into two equally sized edges; the new vertices are projected onto the sphere; and each triangle is replaced by four sub-triangles. The geodesic sphere obtained by subdividing the icosahedron n times is called an n-geodesic sphere. Fig. 3 shows a 0-geodesic sphere and a 1-geodesic sphere.

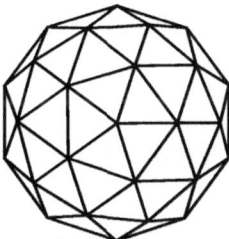

Fig. 3. (Left) a 0-geodesic sphere (icosahedron); (right) a 1-geodesic sphere

The vertices of a geodesic sphere are distributed approximately evenly across the sphere, so the sampling rays are distributed evenly with respect to direction. In practice, most objects to be reconstructed by SFS are composed of block-shaped components, so most rays of the dandelion model intersect the visual hull surfaces with a big incident angle and the intersection computation is well-posed. The rare ill-posed computations can be detected and their results can be abandoned. The geodesic sphere should be subdivided until the rays intersect all connected components of the visual hull. However, the visual hull of an object is usually composed of only one connected component in graphical modeling applications, and thus a 1- or 2-geodesic sphere usually works well.

5   Computing 3D Intersections by Cutting Segments

In this section, we discuss the algorithm for computing all intersections between an arbitrary line/ray/line segment (referred to as the *query line/ray/line segment*) and the visual hull surface, as is required by both initial point computation and triangulation refinement. Because lines and rays can be converted to line segments by first being intersected by a *bounding box* of the visual hull, we focus only on the problem of intersecting a query line segment with the visual hull surface in the sequel. Aganj *et al.* [2] solved this problem by dichotomic search. That approach can only find one of the intersections and is slow and inexact. Liang and Wong [16] solved this problem by traversing the pixels in the image with Bresenham's line algorithm. That approach can only achieve pixel-level precision, which will cause visible artifacts in renderings. We compute all exact intersections efficiently by cutting line segments with the visual hull.

As shown in Fig. 4, for a query segment, AB, we elongate it a bit in both ends to obtain $A'B'$. We use $A'B'$ instead of AB to facilitate processing the end points. Projecting $A'B'$ into each image I, we obtain a 2D line segment $a'b'$. $a'b'$ is then cut with the silhouette represented by oriented polygonal contours. This results in a set of 2D line segments inside the silhouette regions. These 2D line segments are then back-projected into 3D space to produce a set of 3D line segments, which are the result of cutting $A'B'$ with the viewing cone related to I. Next, the intersections of the 3D line segments cut from $A'B'$ by all viewing cones are computed. This gives us the result of cutting $A'B'$ with the visual hull. Finally, each end point of the resulting line segments is checked. If an end point is on AB, it is an intersection between AB and the visual hull surface. The pseudo code of computing 3D intersections is shown below.

```
IntersectWithVHSurf (line_seg AB)
{
    A'B' = Elongate (AB);
    listIntersections = ∅;
    listSegs3D = A'B'; // sorted list of segments in 3D
    for each image {
        a'b' = Project (A'B');
        listSegs2D = SegCutWithSilhouette (a'b');
        listSegs3D' = BackProject(listSegs2D);
        listSegs3D = listSegs3D ∩ listSegs3D';
    }
    for each line segment s in listSegs3D {
        if (s.startPoint ∈ [A, B])
            add s.startPoint to listIntersections;
        if (s.endPoint ∈ [A, B])
            add s.endPoint to listIntersections;
    }
    return listIntersections;
}
```

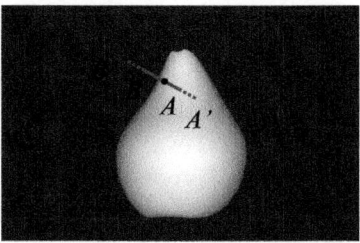

Fig. 4. Computing 3D intersections by cutting line segments with the visual hull

6   Barrel-Grid and Quick Segment Cut in 2D

In the algorithm shown in Sec. 5, segment cut with silhouettes in 2D would be computationally expensive if a brute force algorithm were used. We can notice the characters of the computation:

- the projected line segment is in an arbitrary position;
- the polygonal contour may contain hundreds of line segments and have a complex shape;
- a huge number of 2D segment cuts is involved in the SFS algorithm.

Hence, some special method is needed to make the SFS algorithm efficient. In this section, we present a novel barrel-grid structure and some related algorithms to address this problem.

As shown in Fig. 5, a *barrel grid* divides the image space into a grid of equally-sized square cells; each cell servers as a *barrel* to record the contour segments intersecting the cell. The cells intersecting a line segment s is called the *colored cells* of s.

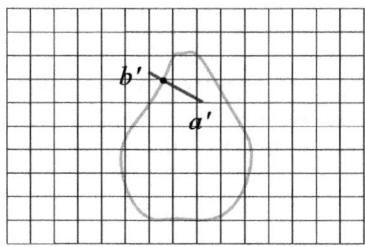

Fig. 5. The barrel-grid structure

The colored cells of s can be computed efficiently with an algorithm similar to the line-drawing algorithm [9] in computer graphics. We address this problem with respect to eight cases, i.e., $i\pi/4 \le \alpha < (i + 1)\pi/4$, for $i = 0, 1, \cdots, 7$, where α is the polar angle $\angle sx$ contained by the oriented line segment s and the x-axis. Here we only discuss the case $0 \le \alpha < \pi/4$, as shown in Fig. 6. Other cases can be processed similarly. The algorithm starts from the *start-cell* recording the start point of s and visits the colored cells column by column from left to right. For the *current-cell* it visits, the algorithm checks the relative position of the *up-right-corner*, UR, with respect to the supporting line l of s. If UR is to the left of l, the algorithm colors (or reports) the current-cell, and set current-cell to the right cell; if UR is exactly on l, the algorithm colors the current-cell and the up and right cells and set current-cell to the up-right cell; if UR is to the right of l, the algorithm colors the current-cell and the up cell, and set current-cell to the up-right cell. The pseudo code is shown below.

```
ColoredCells (line_seg s)
{
    listClrCells = ∅;
    l = SupportingLine (s);
    curCell = Locate (s.start);
    while (curCell.xMin ≤ s.end.x) {
        add curCell to listClrCells; // list of colored cells
        if (curCell.UR is to the left of l)
            curCell = the right cell;
        else if (curCell.UR is exactly on l) {
            if (curCell.yMax ≤ s.end.y)
                Add the up and right cells to listClrCells;
            curCell = the up-right cell;
        }
        else { // curCell.UR is to the right of l
            if (curCell.yMax ≤ s.end.y)
                add the up cell to listClrCells;
            curCell = the up-right cell;
        }
    }
    return listClrCells;
}
```

The barrel-grid structure is initiated by traversing all contour segments, computing their colored cells, and recording the segments in their colored cells. The barrel-grid structure, once initiated, can be used to quickly pick out a few contour segments that

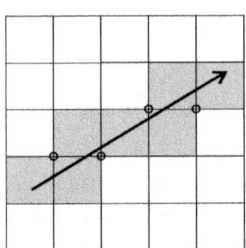

Fig. 6. Colored cell computation

possibly intersect an arbitrary line segment (referred to as a *query segment*) and thus accelerate 2D intersection computations.

For a query segment s to be cut by the silhouette, we first compute the colored cells of s and collect all possibly intersecting contour segments recorded in them. Then the intersections between s and those contour segments are computed. Because all segments are oriented, each intersection can be labeled as either *entry-intersection*, if s makes an obtuse angle with the contour segment's normal n, or *exit-intersection* if s makes an acute angle with n. For example, in Fig. 7 the first intersection between s and the contour is an exit-intersection; the second is an entry-intersection; the third is an exit-intersection. If s and the contour segment are in the same line, an entry-intersection and an exit-intersection are produced. Finally, all intersections, together with the start and end points of s, which can serve both as entry-intersection and exit-intersection, are sorted on s, and a set of segments defined by pairs of *entry-exit-intersections* are produced, which are the result of cutting s by the silhouette.

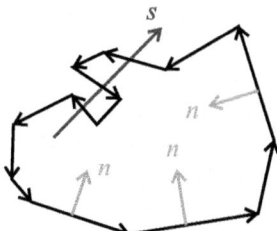

Fig. 7. Cutting line segment in 2D. The silhouette contour is shown in black; the query segment is shown in red; the normals of the contour segments are shown in blue.

7   Refinement Criteria

This section discusses the criteria used to guide the Delaunay refinement, answers the question of what is a *bad facet*, and gives quantitative quality measurements for each bad facet, based on which an order of bad facets can be defined. In the current work, three criteria are taken into consideration with respect to center-center distance, minimum angle, and circumscribing radius, which are similar to those introduced by Rineau and Yvinec [24].

The *center-center distance*, d, of a facet of the restricted DT is the Euclid distance $\|c_1, c_2\|$ between the center of the facet's surface Delaunay ball, c_1, and the center of the facet's circumscribing circle, c_2, as shown in Fig. 8. The *surface Delaunay ball* is a ball circumscribing a triangular facet of the restricted DT and centered at an intersection between the visual hull surface and the Voronoi edge dual to the facet [6]. The quality of a facet with respect to center-center distance is defined as

$$Q_d = min\left\{d_0/d, 1\right\},\qquad(2)$$

where d_0 is a user-designated threshold. A facet is bad if $Q_d < 1$. The center-center distance is used as a measurement of how much a facet deviates from the real visual hull surface, which leads the restricted DT to evolve toward a precise visual hull model.

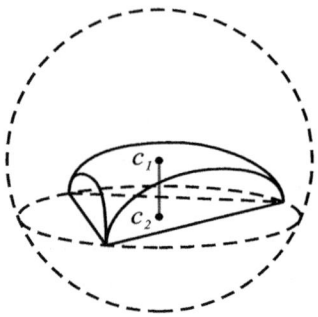

Fig. 8. The center-center distance

The quality of a facet $\triangle ABC$ with respect to the *minimum angle* is defined as

$$Q_a = min\left\{min\left\{\angle A, \angle B, \angle C\right\}/a_0, 1\right\}, \qquad (3)$$

where $\angle A$, $\angle B$, and $\angle C$ are the inner angles of $\triangle ABC$, and a_0 is a user-defined threshold. A facet is bad if $Q_a < 1$. The minimum triangle criterion makes the reconstructed mesh model to be composed of *well-shaped triangles* with minimum inner angles not smaller than a_0, which is favorable for graphical applications.

The quality of a facet with respect to the circumscribing radius r is defined as

$$Q_r = min\left\{r_0/r, 1\right\}, \qquad (4)$$

where r_0 is a user-defined threshold. The circumscribing radius criterion makes the model to be composed of uniformly-sized (not too big) triangles, which is preferred in some graphical applications. We know that the center-center distance is only an approximate measurement of model precision, which may be ineffective in some special cases. We present an example of such special cases in Fig. 9. Limiting the size of the facets can significantly decrease the possibility of the occurrence of such cases.

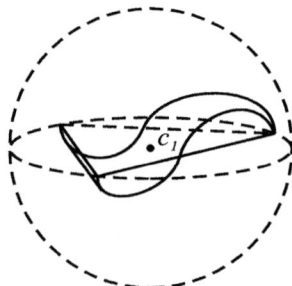

Fig. 9. The center-center distance is zero while the facet deviates much from the wave-shaped surface

The order of bad facets is determined by priority: $f_1 \prec f_2$ if $Q_d(f_1) < Q_d(f_2)$, or $Q_d(f_1) = Q_d(f_2) \wedge Q_a(f_1) < Q_a(f_2)$, or $Q_d(f_1) = Q_d(f_2) \wedge Q_a(f_1) = Q_a(f_2) \wedge Q_r(f_1) \leq Q_r(f_2)$.

The three criteria work together leading the restricted DT to evolve toward a precise, uniformly-sized and well-composed mesh model of the visual hull.

8 Experiments

We tested our algorithm on both synthetic images and real photos and compared our algorithm with both surface-based and volume-based (*volumetric*) approaches. We implemented our algorithm with C++, OpenCV [8] and CGAL [1]. For the surface-based approach, we used the EPVH library (version 1.0.0) [11] for comparisons, which is an implementation of the *Exact Polyhedral Visual Hull* (EPVH) algorithm proposed by Franco and Boyer [12,10]. The volumetric algorithm was implemented based on the techniques presented by Szeliski [25], which use *octree* and *half-distance transform* to accelerate the reconstruction process. A *marching cubes* algorithm [20] was used to extract a mesh model from the voxel model. For all experiments, the input images were rendered or resized to 600×398 pixels. The virtual and real objects are enclosed by a bounding box of $300 \times 300 \times 300$ mm³. For our algorithm, we set the threshold of center-center distance $d_0 = 1$ mm, the threshold of minimum angle $a_0 = 30°$, the threshold of circumscribing radius $r_0 = 20$ mm. For the volumetric algorithm, the maximum resolution of the voxel model was set to 256^3 voxels. For each experiment, we report the number of triangles in the mesh model and the running time, which was measured on a laptop computer equipped with an Intel® Core 2 Duo 2.1 G Hz CPU and 4 Gb main memory.

We first present the results of reconstructing a gnus five tanglecube model from 54 synthetic images in Tab. 1 and Fig. 10. The virtual viewpoints are distributed evenly around the three Cardinal coordinate axes, as shown in the top-right image. From Tab. 1 and Fig. 10 we can see that our algorithm ran moderately fast (40.09s), and reconstructed a model with the highest quality. The efficiency is achieved by the efficient dynamic DT algorithm and our quick 3D intersection computation algorithm. The parameter settings determined that the reconstructed model has a precision as high as 1/300. The model is smooth, as evident in the shaded rendering, because the sampling

Table 1. Numeric results of reconstructing a tanglecube model from 54 synthetic images

	ours	EPVH	volumetric
time (s)	40.09	6.31	147.34
triangles	6,220	8,983	975,544

Table 2. Numeric results of reconstructing a deer model from 36 synthetic images

	ours	EPVH	volumetric
time (s)	14.79	3.73	19.00
triangles	3,693	14,600	173,810

points were computed exactly under well-posed conditions. The mesh model is composed of 6,220 triangles, which is the least among the three algorithms. The vertices are distributed evenly and the triangles are well-shaped, as can be seen from the wireframe rendering, with the minimum angle not smaller than 30°. Such models are very favorable for graphical applications. For example, vertex normals computed as the average of the incident facets' normals produce smooth shading. In comparison, the EPVH algorithm used only 6.31s, which is very fast. However, the reconstructed model has holes on the surface, as can be seen from the shaded rendering. The holes are caused by the essential drawbacks in robustness of surface-based algorithms. The output model is not smooth because some vertices were computed under ill-posed conditions. The mesh model is composed of 8,983 triangles. The vertices are distributed densely around the

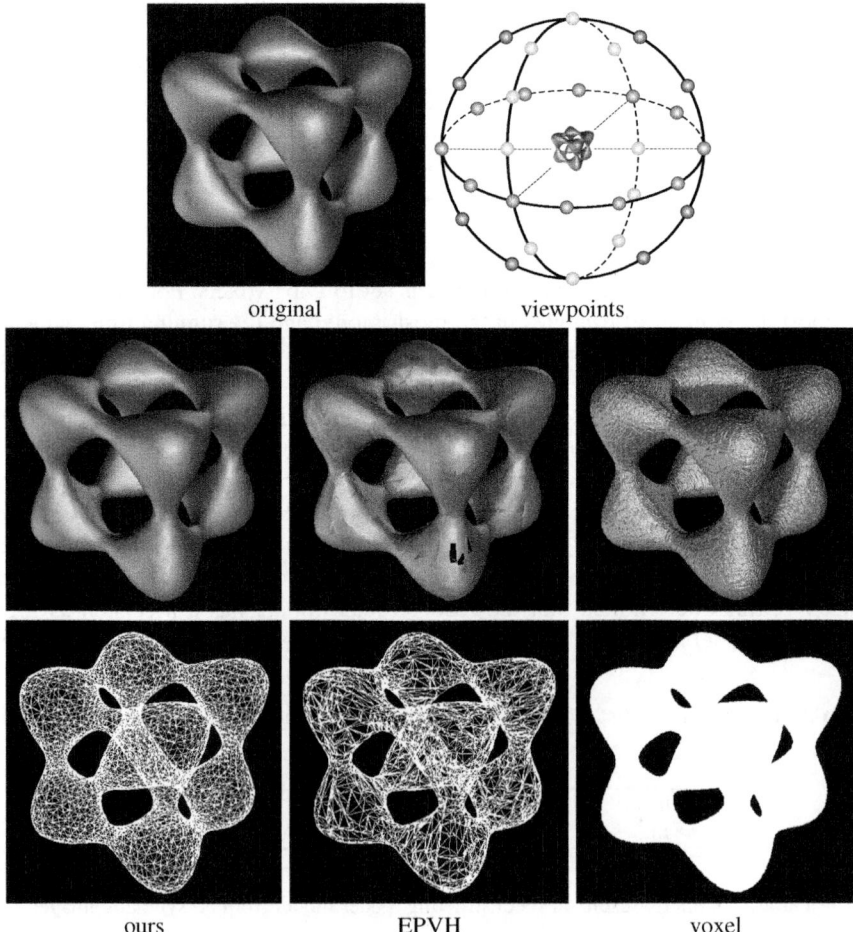

original   viewpoints

ours   EPVH   voxel

Fig. 10. Results of reconstructing a tanglecube model from 54 synthetic images

intersections of the viewing cones, and the triangles are very irregular. Such a model is very disadvantageous for graphical applications. For example, the vertex normals computed as the average of incident facets' normals do not produce smooth shading. The volumetric algorithm is the slowest (147.34 s). The reconstructed model is composed of over-dense vertices. Even with a resolution as high as 256^3 voxels, discretization artifacts are obvious in the shaded rendering.

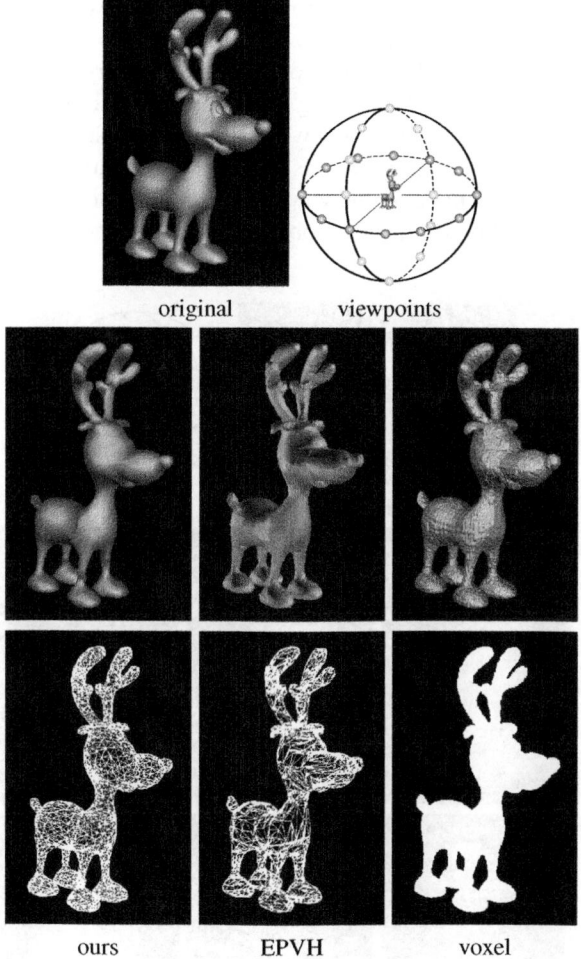

Fig. 11. Results of reconstructing a deer model from 36 synthetic images

Table 3. Numeric results of reconstructing a cat model from 36 real photos

	ours	EPVH	volumetric
time (s)	17.97	1.54	30.00
triangles	3,683	7,880	175,860

We then present the results of reconstructing a complex deer model from 36 synthetic images in Tab. 2 and Fig. 11. The viewpoints were distributed evenly around two of the three Cardinal coordinate axes, as shown in the top-right image. This experiment shows consistent results: our algorithm ran at a moderate speed, and produced the highest quality model. The shaded rendering of the model reconstructed by our algorithm is visually pleasant and very similar to the original model. That is the combining result of precise vertices and well-composed meshing. As can be seen from the wireframe rendering, more vertices of the model reconstructed by our algorithm were used to represent the complex structures. That shows the effectiveness of our adaptive strategy. The EPVH algorithm is the fastest. However some vertices were computed under ill-posed conditions and the mesh model it produced was poorly composed. That causes bad visual effects in the shaded rendering. The volumetric algorithm is the slowest but relatively faster than the first experiment when compared with the other two algorithms. The reasons are: a cubic initial model was used and most part of it was carved away at coarse levels. The output model of the volumetric algorithm is composed of over-dense vertices, and discretization artifacts are obvious. The deer model is pretty complex, so

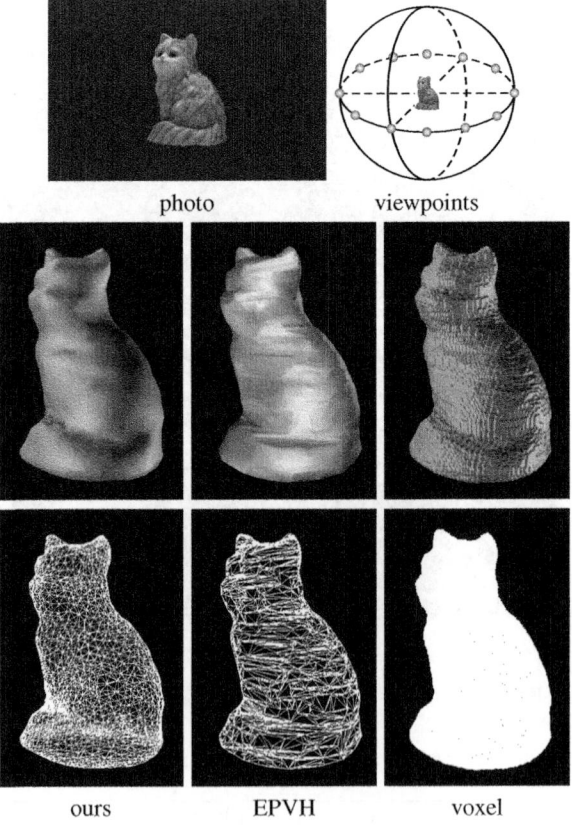

photo   viewpoints

ours   EPVH   voxel

Fig. 12. Results of reconstructing a cat model from 36 real photos

this experiment suggests that our algorithm is able to reconstruct models with complex geometry.

We finally present the results of reconstructing a cat model from 36 real photos. In this experiment, we used an acquisition system composed of a fixed Nikon® D70 camera and a turntable. Both the camera and the turntable were calibrated imperfectly. A plastic cat model was put on the turntable, and 36 photos were taken every $10°$. Figure 12 shows one of them in the top row. The results are shown in Tab. 12 and Fig. 12. Again, our algorithm used moderate time and reconstructed the highest quality model composed of well-shaped triangles. Because the camera and turntable were calibrated imperfectly, this experiment suggests that our algorithm can tolerate sufficient amount of noises and be used in practical applications.

9 Conclusion

In this paper, we have presented a new algorithm for reconstructing high quality visual hull models based on the Delaunay refinement paradigm. The algorithm first computes an initial DT restricted to the visual hull surface and then progressively refines the triangulation by inserting new sampling points, which are the intersections between the visual hull surface and the Voronoi edges dual to triangulation's facets. Because the Voronoi edges are nearly perpendicular to the visual hull surface, computing the intersections is a well-posed problem and thus our algorithm is robust. We compute the intersections by cutting line segments with the visual hull. This problem is then converted to intersecting an arbitrary line segment with polygonal contours in 2D. A new barrel-grid structure is developed for quick intersecting in 2D. Three criteria are used to lead the restricted DT to evolve toward a precise and well-composed mesh model of the visual hull. In conclusion, our algorithm is robust, fast, and fully adaptive, and it produces precise and smooth mesh models composed of well-shaped triangles. The advantages were proved by experiments on synthetic images and real photos.

References

1. Cgal, computational geometry algorithms library, http://www.cgal.org
2. Aganj, E., Pons, J.P., Segonne, F., Keriven, R., Team, W.: Certis: Spatio-temporal shape from silhouette using four-dimensional delaunay meshing. In: IEEE 11th International Conference on Computer Vision. pp. 1–8 (2007)
3. Allard, J., Boyer, E., Franco, J.S., Menier, C., Raffin, B.: Marker-less real time 3d modeling for virtual reality. Immersive Projection Technology (2004)
4. Allard, J., Franco, J.S., Menier, C., Boyer, E., Raffin, B.: The grimage platform: A mixed reality environment for interactions. In: ICVS 2006: Proceedings of the Fourth IEEE International Conference on Computer Vision Systems, p. 46. IEEE Computer Society, Washington (2006)
5. Baumgart, B.: Geometric Modeling for Computer Vision. PhD thesis, Stanford University (1974)
6. Boissonnat, J.D., Oudot, S.: Provably good sampling and meshing of surfaces. Graphical Models 67, 405–451 (2005)

7. Boyer, E., Franco, J.S.: A hybrid approach for computing visual hulls of complex objects (2003)
8. Bradski, G.: Programmer's tool chest: The opencv library. Dr. Dobbs Journal (2000)
9. Bresenham, J.: Algorithm for computer control of a digital plotter. IBM Systerms Journal 4(1), 25–30 (1965)
10. Franco, J., Boyer, E.: Efficient polyhedral modeling from silhouettes. IEEE Transactions on Pattern Analysis and Machine Intelligence (2008)
11. Franco, J.S.: Epvh visual hull library, http://perception.inrialpes.fr/~Franco/EPVH/
12. Franco, J.S., Boyer, E.: Exact polyhedral visual hulls. In: British Machine Vision Conference (BMVC 2003), pp. 329–338 (2003)
13. Isidro, J., Sclaroff, S.: Stochastic refinement of the visual hull to satisfy photometric and silhouette consistency constraints (2003)
14. Kutulakos, K., Seitz, S.: A theory of shape by space carving. International Journal of Computer Vision 38(3), 199–218 (2000)
15. Laurentini, A.: The visual hull concept for silhouette-based image understanding. IEEE Transactions on Pattern Analysis and Machine Intelligence 16(2), 150–162 (1994)
16. Liang, C., Wong, K.Y.K.: Exact visual hull from marching cubes. In: VISAPP, vol. 2, pp. 597–604 (2008)
17. Liu, X., Yao, H., Chen, X., Gao, W.: Visual hull embossment by graph cuts. In: 2006 IEEE International Conference on Image Processing, pp. 2205–2208 (2006)
18. Liu, X., Yao, H., Chen, X., Gao, W.: Shape from silhouettes based on a centripetal pentahedron model. Graphical Models (2008)
19. Liu, X., Yao, H., Gao, W.: Shape from silhouette outlines using an adaptive dandelion model. Computer Vision and Image Understanding 105(2), 121–130 (2007)
20. Lorensen, W.E., Cline, H.E.: Marching cubes: A high resolution 3d surface construction algorithm. In: Proceedings of the 14th annual conference on Computer graphics and interactive techniques (1987)
21. Martin, W., Aggarwal, J.: Volumetric description of objects from multiple views. IEEE Trans. on PAMI 5(2), 150–158 (1983)
22. Matusik, W., Buehler, C., McMillan, L.: Polyhedral visual hulls for real-time rendering. In: The Eurographics Workshop, London, United Kingdom, pp. 115–126 (2001)
23. Mulayim, A.Y., Ozun, O., Atalay, V., Schmitt, F.: On the silhouette based 3d reconstruction and initial bounding cube estimation (2000)
24. Rineau, L., Yvinec, M.: A generic software design for delaunay refinement meshing. Computational Geometry: Theory and Applications 38(1-2), 100–110 (2007)
25. Szeliski, R.: Real-time octree generation from rotating objects. Tech. rep., Cambridge Research Laboratory, Digital Equipment Corporation (1990)

Geosimulation of Geographic Dynamics Based on Voronoi Diagram

Mir Abolfazl Mostafavi, Leila Hashemi Beni, and Karine Hins Mallet

Center for Research in Geomatics, Department of Geomatics, Laval University
Quebec City, G1V 0A6, Quebec, Canada
Mir-Abolfazl.Mostafavi@scg.ulaval.ca,
{Leila.hashemi.1,karine.hins-mallet.1}@ulaval.ca

Abstract. Geographic space is typically conceptualized either as discrete objects or as continuous fields. Considerable efforts have been carried out for the representation and management of the spatial data, based on the object view of space. However, field-based data models are less developed in GIS, especially when it comes to the modeling and representation of their spatial dynamics and behaviors. The limitations of GIS are mostly related to the 2D and static nature of their spatial data structures. In this paper, we explore the potentials of Voronoi diagram as an alternative spatial data structure that can allow more realistic representation of geographic dynamics, especially of dynamic fields and processes in 2D and 3D spaces. The paper presents a review on how different types of Voronoi diagrams for points, line segments and polygons could be effectively used to represent and simulate the geographic dynamics in different contexts.

Keywords: Voronoi diagram, spatial process, dynamic fields, behavior, simulation, cellular automata, geographic dynamics.

1 Introduction

Representation of dynamic fields and processes is a key issue for geographic information scientists. Most spatial phenomena are dynamic. What is intended by 'dynamics' here are changes through time and space. Global warming, global outbreak of H1N1 Swine Flu, inundation, erosion, land cover change, urban growth, and traffic and pedestrians' behavior are examples of dynamic spatial phenomena in the geographic domain. A realistic modeling and representation of such changes is necessary for better characterizing, understanding, analyzing and predicting the behavior of those geographic dynamics [3], [31]. Reference [33] argues that the domain of geographic dynamics is very vast, that its representation is a concern for many disciplines, and that a wide range of tools have been developed for this purpose. MODFLOW (USGS) for ground water modeling, POM (Princeton Ocean Model) for oceanographic modeling [35], SLEUTH for urban growth [34], and agent based models for pedestrians' behavior [37] are a few examples of those tools. While these simulation tools are

M.L. Gavrilova et al. (Eds.): Trans. on Comput. Sci. IX, LNCS 6290, pp. 183–201, 2010.
© Springer-Verlag Berlin Heidelberg 2010

relatively efficient for description of spatial dynamics in different contexts, they generally suffer from insufficient spatial analytical components.

In Geographic Information Systems (GIS), considerable effort has been made to represent and manage spatial data based on the object view of the space. However, field-based data models are less developed in GIS, especially when it comes to the modeling and representation of geographic dynamics. GIS have been criticized for not being able to provide the necessary functionalities for effectively representing, analyzing and predicting the behavior of those geographic phenomena. This is because current GIS data structures are mostly 2D and static, not being well-defined to represent geographic dynamics PCRaster [36] is one of the tools in GIS sciences that allows modeling of a dynamic process as a sequence of snapshot raster layers and which includes a range of functions for spatiotemporal environmental modeling in 2D and 2.5D. However, geographic dynamics are far more complex and their representation and visualization need more investigation. In particular, Geographic dynamics are multi-dimensional and need to be explored in multi-resolution and multi-scale representations. Therefore, understanding of geographic dynamics necessitates more comprehensive methods and tools to address their behavior, interactions, and relationships as well as their evolutions in time and space [31], [32].

The purpose of this paper is to explore the potentials of the Voronoi diagrams as an alternative spatial data model allowing for more realistic representation of the geographic dynamics in 2D and 3D spaces. To this end, the paper presents how different types of Voronoi diagrams for points, line segments and polygons could effectively be used in different contexts of application to represent and simulate dynamic phenomena and their behavior.

The remainder of this paper is organized as follows. In section 2, we describe conceptual approaches for modeling of geographic dynamics in GIS. This will be followed by a description in section 3 of the methods used to describe the dynamic behavior of spatial phenomena. Section 4 deals with the role of the Voronoi diagram in simulation methods, by presenting several case studies. Finally, we present some results and discuss the potentials and limitations of the Voronoi diagram for the simulation of dynamic fields and processes and then we present the conclusions of the work.

2   Modeling Geographic Dynamics in GIS

In geographic information sciences, fields and processes, which are usually characterized by geographic dynamics, are defined as spatial continuous phenomena. The value of those phenomena depends on their locations [4], [5]. Topographic elevation, air temperature and precipitations are examples of spatial fields. In reference [6], a 2D field (F) is defined as any single-valued function of location in 2D space: $F= f (x, y)$. This definition can be generalized to 3D as: $F= f (x, y, z)$. This value becomes also a function of time when dealing with a dynamic field or process. Depending on how time is taken into account in the function, two different approaches can be introduced: We can investigate the changes of a field at a fixed location over time: $F(t) = f(x, y, z, t)$; or we can describe the changes of the field at a time-depended location:

$$F(t) = f(x(t), y(t), z(t)) \tag{1}$$

Dynamic fields generally have a very strong spatial component and researchers in the GIS field have been interested in modeling and representing them within GIS. Traditionally, simulation tools have been developed outside GIS. However, these tools suffer from an insufficient spatial analytical component, a low performance of the spatial data visualization capability, and non-conveniences for spatial data integration such as digital maps, satellite image, and aerial photos [7]. Therefore, the integration of simulation tools and GIS was considered by many researchers, the benefits of which were confirmed by several papers published in this area [8], [9], and [10].

Depending on the level of interaction between a simulation tool and GIS, three integration approaches can be distinguished. One is the Loose Coupling approach, which is defined as merely passing input and output between a GIS and a simulation tool. In this level of integration, the input into simulation tools comes from GIS and the output of modeling is exported back to GIS for analysis and visualization. Next is the Tight Coupling approach, which involves data some processing in GIS. In this approach, the process model is still made using simulation tools outside of a GIS, however, data are exchanged automatically between those tools. Finally, the Full Coupling approach is defined as embedding a process model in a GIS. It involves implementing a process model within the GIS and taking full advantage of the built-in GIS functionalities. In addition, this allows a GIS to go beyond being a simple data management tool and to offer more sophisticated analyses and simulation capabilities for representation of natural phenomena. This later approach is referred to as a GIS-based simulation tool. Current GIS do not support simulation of dynamic fields as their structures are generally 2D and static [25].

Dynamic fields are continuous, making it practically impossible to measure them anywhere and anytime. As a result, our knowledge of spatiotemporal fields is usually limited to a set of observations in a finite set of locations and time instances. These data are usually composed of a set of unconnected points. Each point is defined by its position in the 2D or 3D spaces and by its attribute (field value) at a given time.

To represent a continuous phenomenon from a set of discrete points, it is necessary to create a tessellation that rebuilds the continuity and connectivity between the discrete observations. A tessellation also referred to as mesh or grid is a partition of the space by a set of elements such that the union of all elements completely fills the space [2]. In GIS, according to the characteristics of the studied phenomena and the data acquisition methods, there exist six commonly used methods for the representation of continuous fields [1]. These include raster (fig. 1a), polygons (fig. 1b), TIN (fig. 1c), irregular point sets (fig. 1d), regular point sets or grids (fig. 1e) and digitized contour lines (fig. 1f).

As one can see from the Fig. 1, however, not all those methods allow for a continuous representation of a field. In addition, it is very difficult to manage the temporal aspect of dynamic fields and processes using those representations.

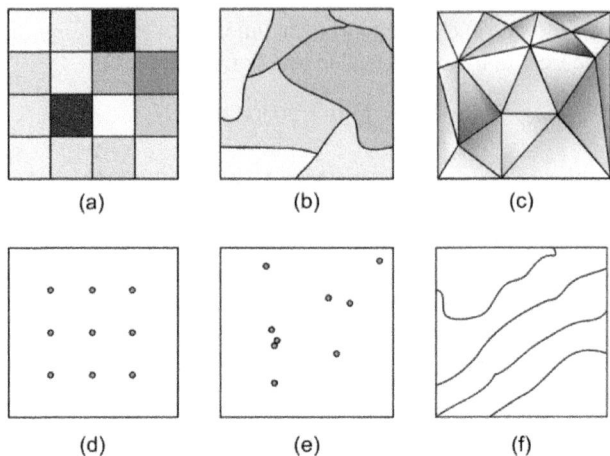

Fig. 1. Six different methods for the spatial representation of a field in 2D, a) raster, b) polygons, c) TIN, d) irregular point sets, e) regular point sets or grids, and f) digitized contour lines

Methods that can provide a continuous representation of a field are referred to as 'complete representations' and can be of two types; regular and irregular meshes. The elements of a regular mesh have uniform shapes and sizes such as 2D rectangles and pixels or 3D cubes and vexes, with implicit topology between those elements. However, a major limitation of regular meshes is related to the handling of field data (unconnected points) with an irregular distribution. In this case, a large number of elements are required for a fine resolution and representation of the field. To overcome the limitations of regular meshes, hierarchical data structures can be used. Quadtrees and Octrees are two examples of these data structures in 2D and 3D spaces, respectively. These structures subdivide the space into four squares (Quad tree, in 2D) or eight cubes (Octree, in 3D) of equal size so that either each element would contain one homogeneous region or reach a given resolution. Although these methods required less memory for the storage of a fine resolution mesh, their trees may be unbalanced for an irregular data distribution [11]. In addition, a small change in the field data may result in quite different trees.

Irregular meshes can also be used to model a continuous phenomenon. The elements of irregular meshes can be of any size and shape such as triangles and polygons in 2D or tetrahedrons and polyhedrons in 3D. Therefore, they can be adapted to the point distribution and provide conformity to complex data distribution. However, in irregular meshes, the topology needs to be computed and stored explicitly and updated after any change, which can be a difficult task for current GIS data structures. The data structures are static, and thus not well-adapted for the management of topology changes [25]. Therefore, following any change in the spatial information, all spatial relationships must be rebuilt. In addition, the GIS data structures are not capable of representing both objects and fields at the same time as required in some simulation applications. Tracking contaminants in groundwater flow is an example of this requirement.

References [14] and [13] suggest that the Voronoi diagram and its dual Delaunay triangulation can be a good choice for the representation and simulation of dynamic fields within GIS. In this paper, we present some interesting results that we have obtained from our ongoing research on the employment of the dynamic and kinetic Voronoi diagrams for simulation purposes in the context of geographic process representation and simulation. It should be mentioned that the Voronoi diagram has a broad application in other contexts, the exhaustive review of which is beyond the objective and scope of this paper. In the next section, we explain how the dynamic behavior of a field is described, and how Voronoi diagram can be used to represent it.

3   Representation and Simulation of the Behavior of a Spatial Dynamic Phenomena

Depending on the nature of a spatial field, different approaches are used to describe its dynamic behavior. In some cases, dynamic behavior of a field is described using a set of partial differential equations (PDEs). Other types of descriptions can as well be seen in the literature for spatial dynamic processes based on a set of rules. The type of the description of a field's dynamic behavior determines the nature of the solutions for its modeling, representation and prediction.

3.1   Numerical Integration Methods for Simulation of Dynamic Fields

There are two fundamental methods typically used to numerically integrate the PDEs that describe the dynamic behavior of a spatial field; namely, the Eulerian and the Lagrangian methods (Fig. 2). Eulerian methods describe changes of the field values as they occur at a fixed location. A high resolution tessellation of the simulation domain is therefore required in order to properly take into account the spatial variation of the field. In Lagrangian methods, the simulation domain is discretized to a set of moving particles and the behavior of a dynamic field is described following the moving particles along their trajectory. The later methods are often considered more precise as the numerical errors are less important [17]. However, Lagrangian methods have their own limitations as the spatial tessellation supporting these methods (ex. TIN) may undergo very large distortions, which compromises their suitability for accurate representation of the behavior of a dynamic field. To overcome this limitation, Free-Lagrangian methods are used, where particles are allowed to change their neighborhood relations. This property allows for the representation of complex behavior of dynamic fields without the problems of mesh tangling. However, Free-Lagrangian methods are more complex because the spatial data structures that are used to manage topological

Eulerian
approach

Lagrangian
approach

Free-Lagrange
approach

Fig. 2. Different methods for the numerical integration of dynamic fields

changes in the mesh should support local changes, which is not the case in the most existing methods [14]. Dynamic and kinetic spatial data structures developed based on Voronoi diagram could be an alternative solution to these problems. This will be discussed in detail further on in this paper.

3.2 Geo-simulation Methods for Representation and Management of Spatial Dynamic Processes

As mentioned in the previous section, not all the dynamic spatial phenomena are described by partial differential equations. Urban dynamics such as those of the traffic, of people displacement, or of the behavior of a group of people following an emergency situation, are usually described by a set of rules that define different interactions between those people (i.e. agents) involved in those events as well as their interactions with the physical environment. Different approaches are used to simulate these complex dynamic phenomena, which are usually referred to as geosimulation methods [12] and [18]. Cellular automata approaches are among the geosimulation methods used for the simulation of very complex dynamic spatial phenomena. In these approaches, the complex behavior of a dynamic phenomenon is represented by the integration of more simple and local interactions between simple components of the phenomenon. For this purpose, the initial state of each simple component (cell) is defined and its state computed using a set of transition rules during a given period of time. These rules describe the nature of the interactions between each cell and its neighbors during the simulation process.

Most of the cellular automata methods are based on regular tessellations [28], [29], and [30]. A regular tessellation discretizes the space into cells of the same forms and sizes [2]. While rectangular tessellations are more frequent in geosimulation domain, however, triangular and hexagonal tessellations are also used for this purpose (Fig. 3). Here, the spatial resolution of a tessellation depends on the precision required for the representation of a spatial phenomenon, which usually remains constant throughout the simulation domain. Adjacency relations are one of the most important types of

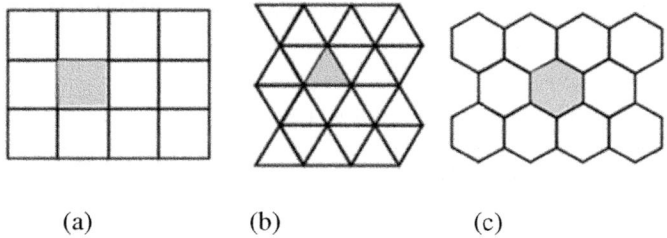

(a)   (b)   (c)

Fig. 3. Different regular tessellation types used in cellular automata approaches

information required in cellular automata methods. In a regular tessellation, the number of the adjacent cells is the same for a given cell, which usually includes only the cells that share their border with the central cell.

Another important point is that complex dynamic phenomena can be studied on different macro, meso and micro scales. As a matter of fact, a specific pattern of a dynamic phenomenon can only appear in a specific spatial resolution. It is therefore important for the geometrical tessellation, which supports the cellular automata based simulation, to not only have the capability of being adapted to the reality of the phenomenon but also of supporting the aggregation of the lower scale interactions to higher levels. This will allow the simulation system to represent the phenomena in other meso and macro scales, and will facilitate its characterization and understanding. Current cellular automata methods are mostly based on regular and static grids in which the generation of hierarchical aspects of the simulation is relatively easy. There are also other methods based on the triangular and hexagonal tessellations with different degrees of detail (Fig. 4). However, the inadaptability of those structures at different levels of the aggregation to the real world phenomena is still a problem [12, 26].

Voronoi diagrams can be considered as an alternative tessellation to improve the adaptability of these methods to the reality. However, their application as a hierarchical structure is still in its starting stage. This will be discussed in more detail in the coming sections.

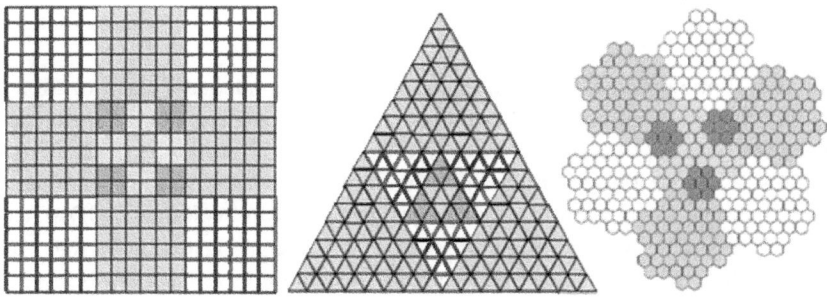

Fig. 4. Hierarchical rectangular, triangular and hexagonal tessellations used in geosimulation

4   Voronoi Diagram as an Alternative Tessellation for Geographic Dynamics Representation

In this section, we introduce the Voronoi diagram as an alternative data model that can effectively support a more realistic representation of a dynamic spatial phenomenon. We first present a brief definition of the Voronoi diagram and then illustrate its adoption to the simulation of spatial dynamic phenomena for both categories of the geographic dynamics that were described in the previous sections. Then, we will discuss the potentials and limitations of this diagram in the context of spatial simulation.

4.1   Definition of Voronoi Diagram

Given a set of a finite number of points in the Euclidean plane, we associate each location in the space with its closest member(s) of the point set with respect to the Euclidean distance. The result is a tessellation of the plane into a set of regions associated with individual members of the point set. This tessellation is called the "ordinary planar Voronoi diagram", generated by the point set. The regions constituting the Voronoi diagram are referred to as Voronoi cells [19] (Fig 5a).

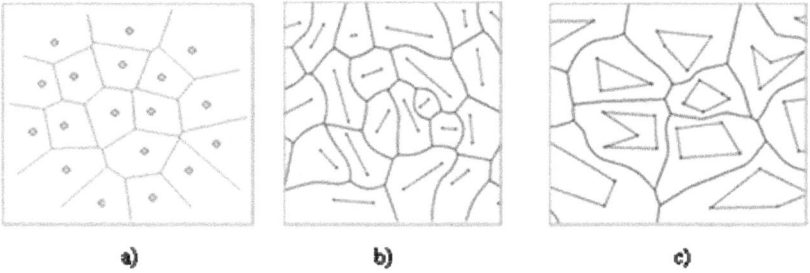

Fig. 5. The Voronoi diagram for a set of a) points, b) line segments, and c) polygons in the Euclidean plane

The useful properties of Voronoi diagram provide an adequate discretization of the space for simulation purposes, where the spatial proximity is a key issue for defining the local interactions of the basic elements (cells) used in such a process. This is applied not only to the Voronoi diagram of a set of points but also to other types of the Voronoi diagram; i.e. those for line segments, polygons and volumes in 3D space [14]. Dynamic and kinetic Voronoi diagrams offer very interesting functions that could be used in simulation of a dynamic phenomenon. It is possible to have on-the-fly interactions with the objects in the simulation domain and insert a constraint, or remove an object from, that domain and study the overall behavior of the dynamic phenomena.

In the following section, we describe the potential of the Voronoi diagram to support Eulerian and Lagrangian based simulation approaches as well as its application to the geosimulation methods based-on cellular automata.

4.2   Voronoi Diagram as an Underlying Mesh for Eulerian and Lagrangian Methods of Simulating 2D and 3D Dynamic Fields

As mentioned in the previous sections, both Eulerian and Lagrangian methods rely on a geometric tessellation for the simulation process called underlying mesh [17]. The role of the tessellation is initially to discretize the domain of simulation to a set of basic elements called particles. Based on these elements, a set of discrete equations are then developed that define the interaction of each element with its neighbors.

As an underlying mesh, not only can the Voronoi diagram adapt itself to an irregular distribution of a set of points sampled from a field, but it can also readily and clearly define the spatial relations between those points, which are also called the generators of the cells in a Voronoi tessellation (Fig. 6). Each cell can have an arbitrary number of neighbors and is retrievable when needed.

In addition, dynamic operations such as local insertion and deletion in a dynamic Voronoi diagram and point movement in a kinetic Voronoi diagram can have a great utility for the representation of the behavior of a dynamic field. These operations allow local edition and modification of the mesh, which is usually necessary for the local refinement and update of the tessellation without having to rebuild the whole structure.

Regarding the Eulerian approaches, several works have been presented in recent years using VD and its dual DT as the underlying mesh for dynamic field representation. Authors of [21] have applied DT and VD to a reservoir simulation using 3D seismic images and have demonstrated the potentials of both DT and VD for the simulation of a fluid dynamics. Reference [20] applied these data structures to a groundwater simulation in 3D space and showed that VD is well-adapted to the Control Volume Finite Element (CVFE) method. The CVFE methods are based on the principle of mass conservation, whereby, a volume of influence is assigned to each point and a set of equations are used to describe the interaction of each element with its neighbors. This interaction is expressed by the mass balance, which states that the difference between inflow to, and outflow from, each element must be equal to the variation in the fluid stored in the same volume.

Kinetic Voronoi diagram has a significant potential for the representation of geographic dynamics as it can effectively deal with moving objects in the space. Kinetic Voronoi diagram can also be used as a very strong means of developing efficient simulation tools based on Lagrangian methods. As we mentioned earlier in this paper, in Lagrangian methods, particles move and interact with each other and it is the result of these interactions that describes the whole behavior of a dynamic field. Such a movement could be supported by a kinetic Voronoi diagram. Depending on the level of the irregularity of the behavior of a field, the efficiency of the Lagrangian approach could vary significantly. If the motion is smooth and there is no necessity for an extensive change in the topological relation between the moving particles, the method will be efficient enough. Otherwise, we would need further improvement in the supporting data structure of the method. This will be described in the next section.

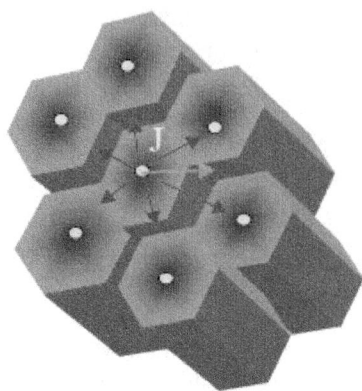

Fig. 6. Particles and their interactions as defined by a Voronoi diagram (redline are the forces describing the interactions between the central cell and its neighbors and the orange line is the resultant force)

4.3   A Free Lagrangian Method Based on the Kinetic Voronoi Diagram Applied to Global tide Simulation

In Lagrangian Methods, with a very irregular behavior of a dynamic field, we need to change the connectivity relation between particles. Otherwise, we will have a mesh tangling problem, which would interfere with the proper execution of the simulation process. Kinetic Voronoi diagram could efficiently support these modifications during the movement of the particles. Based on this idea, extended Lagrangian methods have been developed that are commonly called Free-Lagrangian methods. In these methods, following the topological changes in the mesh, we need to update the physical parameters of the affected points. Governing equations that define the nature of the dynamic field behavior are used to assign updated physical parameters (such as velocity, acceleration, volume, etc.) to each moving particle and its neighbors.

In order to illustrate the application of the kinetic Voronoi diagram for the simulation of a 2D dynamic field, we present the global tide simulation based on this data structure on the globe. This example uses a discrete form of a governing equation that describes the behavior of the global tide. As the first step, the fluid region is sampled by a discrete set of points (particles) and a Voronoi-based mesh is used to tessellate the surface of the oceans on the globe. Next, an initial velocity and a height are assigned to each element of the mesh. The law of 'conservation of mass' is what inspires the basic assumption here. This law indicates that the mass of each element must be constant during the simulation process. This means that, if the surface is changed for each water column, then the height must follow this change. The height of each particle is computed by: $Hi = Mi/Ai$; where Mi and Ai are the mass and area, respectively, of the ith particle. Fig. 7 illustrates the relation of the height of a water column to the forces that are exercised on a given particle.

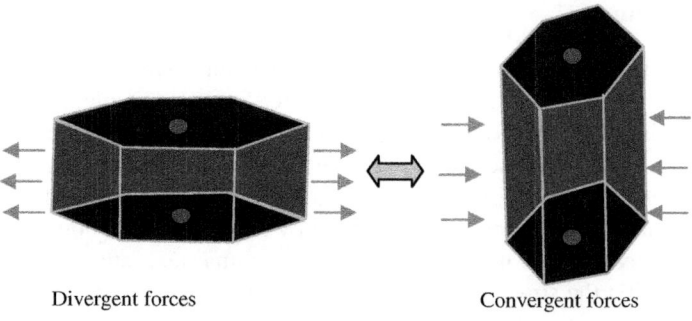

Divergent forces   Convergent forces

Fig. 7. Estimation of height of a water column based on the law of mass conservation

The motion of each particle is the result of its interactions with its neighboring elements. The kinetic Voronoi diagram provides an event-driven method for the simulation of the global tides. This is due to the fact that the motion of the individual particles within the simulation domain is based on a priority criterion. This criterion is defined in terms of the answer to the question 'where will the next topological event happen in the simulation domain?'. In practice, therefore, in order to facilitate the management of the priority of the particle, a priority queue is used which should be constantly updated following each topological event in the mesh.

Fig. 8 illustrates the results obtained from the simulation of global tides for a given period of time. We can see in the figure that the initial area of the particles has changed during the simulation process. As illustrated in Fig 7, a larger area describes a low water tide, and a smaller cell indicates the existence of high water in the region.

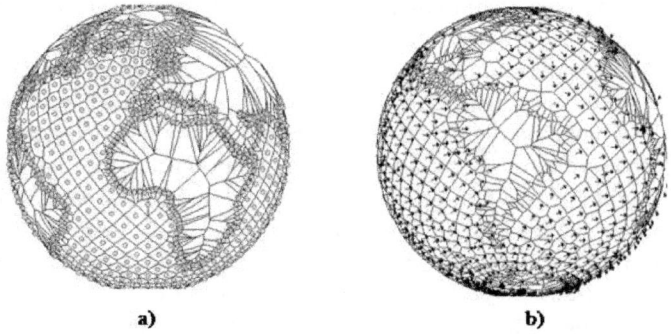

a)   b)

Fig. 8. Simulation of global tides using kinetic Voronoi diagram: a) initial state, b) results after a given simulation time

4.4   3D Kinetic Voronoi Diagram for the Simulation of a 3D Gas Expansion Problem

The second example consists of a 3D hydrodynamics simulation case study which is based on a 3D kinetic Voronoi data structure. In this example, we have investigated the potential of the 3D kinetic Voronoi diagram for a gas dynamics simulation. The governing equations describe mainly the conservation of the mass, momentum and the total energy for a compressible fluid flow [22]. For the sack of simplicity, we consider that the fluid domain is bounded by 3D rectangular boundaries, and that the fluid can move only in the x direction. For the numerical integration of the governing equations, first the simulation domain is discretized using a 3D Voronoi diagram with each Voronoi polyhedron representing a flow particle. Therefore, the relationship between mesh cells is explicitly stored and automatically maintained during the simulation process. To start the simulation, the values for initial velocity (Vx, Vy, Vz), pressure (p), density (ρ), internal energy (e), and mass (m) are assigned to each cell (Fig. 9). In this example, we have assumed that the middle of fluid domain has a higher density and pressure, and that, according to the law of the conservation of mass, the mass of each particle remains constant during the simulation process. Each polyhedral cell shares a face with each of its neighboring cells. The surface of those faces as well as the volume and the shape of the Voronoi polyhedron may change during the simulation process (Fig.9). Similar to the 2D algorithm, in the 3D space, the first topological event for each cell is computed and stored in a priority queue. Then, the first particle on the priority queue is moved to its new position using a velocity vector. The velocity vector is the result of the total force acting on the cell by its neighboring particles. The forces of the neighboring particles are computed on the polyhedron face. Following the movement, the shape and physical attributes (quantities) of the particle and its neighbors are updated. The shape and topological information are updated following a flip23 and flip32, and the physical attributes are computed using the governing equations. Consequently, for each polyhedron, the number of the faces and their respective area change in the new position. Following this step, the next closest topological events for the particle and its neighbors are computed and the priority queue is updated for these new values. This procedure is repeated until the equilibrium is created between the pressures of particles.

The results of this example show that the particles with higher pressure in the middle part of the simulation domain begin to move outward symmetrically and that they become larger in volume. According to the assumption that the mass of particles is constant during the simulation process, when the volume of a particle increases, the density as well as the pressure of the particle decrease. This process continues as long as the equilibrium does not exist in the simulation domain. Results also suggest that the proposed algorithm properly manages the movement of the particles in 3D space, and that the dynamic behavior of the gas flow resulting from the simulation process corresponds to our expectation. This further implies that the gas flow completely fills the simulation domain and that equilibrium is achieved. The initial results obtained in this case study are very promising as the data structure is maintained during the

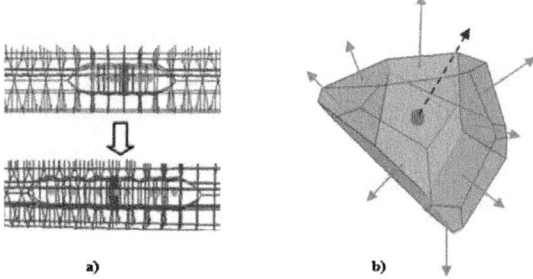

Fig. 9. a) Simulation of gas dynamics using 3D kinetic VD, b) a Voronoi cell and it interactions with its neighbors; the arrows represent the local forces that each moving voronoi polyhedronacts on its nearest neighbours

simulation process representing correctly the geometrical and topological information of the underlying mesh. The results of the simulation are reasonably comparable to the one obtained from the analytical solution of the governing equations.

4.5   Geosimulation Based on Cellular Automata Using Voronoi Diagram for Urban Dynamics Studies

As mentioned in previous sections, geosimulation based on cellular automata typically uses a regular grid for representation and simulation of complex dynamic phenomena [12]. Cellular automata based methods can not only represent a dynamic phenomenon in a given scale but can also support its representation at different levels of detail using a hierarchical data structure. The latter structure allows the aggregation of the lower level information to represent the behavior of a dynamic phenomenon in meso (less detailed level compared to micro level) or even macro (less detailed level compared to meso level) spatial levels. However, the use of regular grids poses important limitations to an accurate geosimulation process at any level of detail ([26], [27]). The real world spatial phenomena are inherently too complex for the regular grids to be adapted adequately to the reality of the simulation domain. This results in a less accurate representation of the real world phenomena and of their dynamics and interactions.

Voronoi diagram presents an interesting alternative model that can support the geosimulation methods by providing an adaptive grid for cellular automata applications. The model is capable of adapting itself to the complex reality and of properly defining the adjacency relationships necessary for the definition of the interactions between elements of the cellular automata. Furthermore, geosimulation applications are generally designed for decision making purposes where we need to produce interactively several scenarios for the resolution of a problem and choose the best solution. In this context, the local dynamic and interactive operations in Voronoi diagram can be very interesting and helpful. As they provide the possibility of local inserting, deleting and moving of objects within the simulation domain. For example, one may be interested

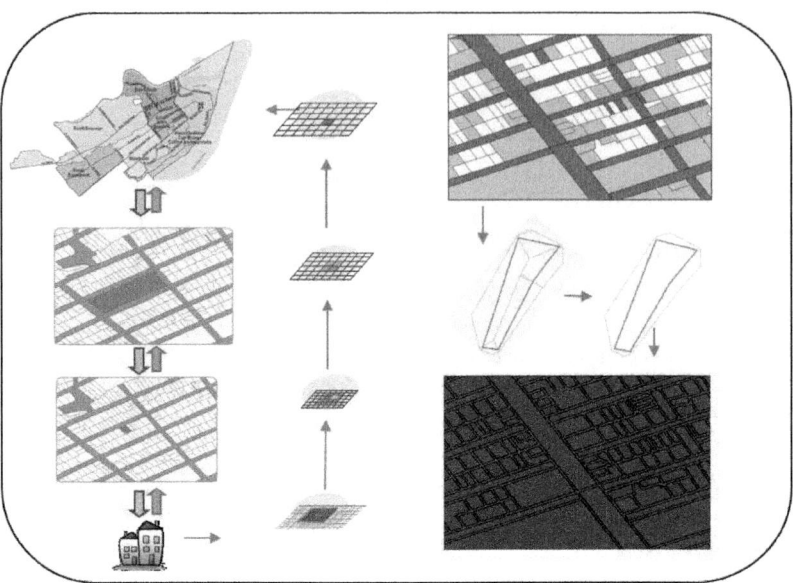

Fig. 10. Reconstruction of a cadastral topological map based on Voronoi diagram for line segment and polygons

in analyzing the urban dynamics in the presence or absence of a shopping center in a specific location in the city. Using the local dynamic operations enables one to create and insert these new objects or remove an existing one efficiently and on-the-fly.

Furthermore, Voronoi diagram can be used to support geosimulation applications at different levels of detail. Some initial work has been done on the Voronoi hierarchies [23]. However, to efficiently support very complex hierarchical geosimulation applications, further research and development are needed.

Figure 10 shows the application of the Voronoi diagram to line segments and polygons for urban dynamics simulation. Here, the Voronoi diagram for line segments and polygons is used as an underlying grid for a gentrification application in a district in Quebec City. Through gentrification process, local authorities in the city have tried to revitalize and help the economic development of the district. In this application, we were interested in studying the dynamics of the house prices, inoccupation level of the residents, and the education level and income of the population in the district. It was assumed that a fiscal exemption for companies would encourage them to move into the area and invest in the specified district of the city, which would in turn have a significant impact on the population dynamics as well as on the real-estate market. For this purpose, we had access to different types of data including the cadastral data of the city and the land use data as well as the data on the population (e.g. age, income, education, job, etc.). Different information on houses such as their price, area, position and their accessibility to the city centre was also available (Fig. 11).

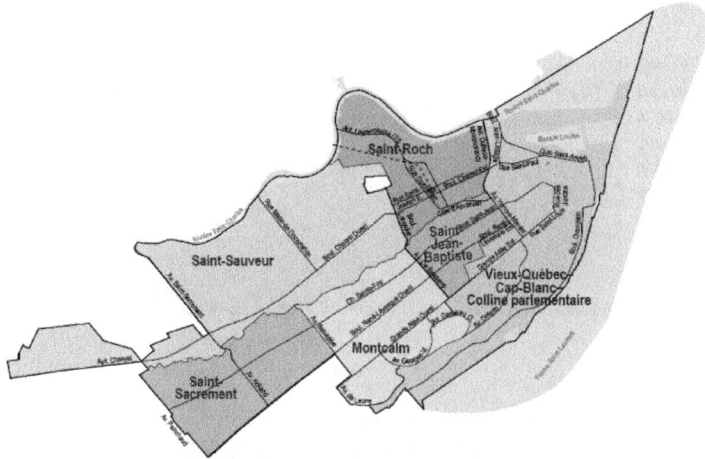

Fig. 11. The district of the "La Cité" in Quebec City, where a gentrification process has been applied

The simulation process was carried out over a 50-year time period with the values being updated at 3 month intervals. The geosimulation process was performed in several steps as follows:

1. In the first step, the simulation process was initialized. This included the identification of the values of all the houses based on the factors that influence the price of the residences such as proximity to a shopping center or a school. The level of education, monthly rental prices and family incomes were also among the initial information used in the simulation process.

2. Next, an iterative process was set up whereby information was gathered on agents (i.e. people and families) looking for a house to buy or rent with respect to their criteria. Following this step, and depending on the level of the demands for each house or an apartment, its price or its monthly rental price was updated.

3. In the final step of the simulation, the mean prices for purchase or rent in the district and in the municipality level were updated by means of an aggregation process applied to the local values in the micro level. These updates also included other information such as the income of each family and the education level of the people in the district.

Fig. 12 shows some results obtained from the simulation process, which describe the impacts of the gentrification process on the monthly rental prices of the houses and also on the education level of the people living in the district of St-Roch in Quebec City during the simulation period. The results indicate that the request for the available residents increased more dramatically in this district than in the neighboring ones. The results are also suggestive of an increase in the education level as well as in the income of the people living in the district as compared to the corresponding values at the initial state; that is, before the simulation process.

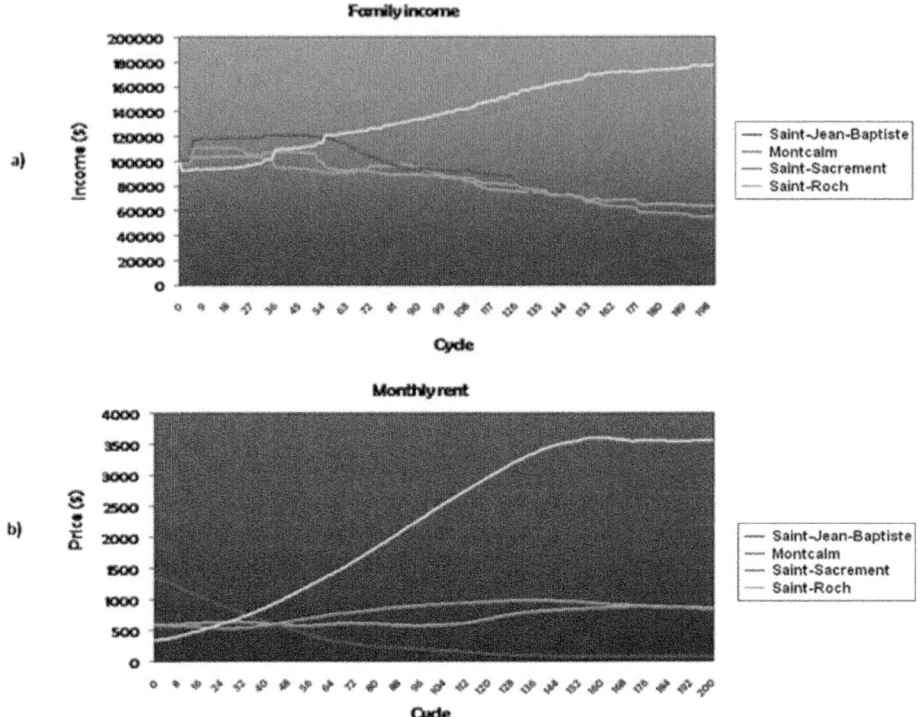

Fig. 12. The impact of the gentrification on a) family income, and b) the monthly house rental prices, in the district of St- Roch

Finally, as we can see in figure 12b that the house rental prices in the district increased dramatically compared to their initial values at the start of the simulation process.

5   Discussion and Conclusions

This paper is a review on the potentials of the Voronoi diagram for the representation and simulation of geographic dynamics. The paper suggests Voronoi diagram as a very powerful spatial model that offers several advantages over other existing spatial models in the representation of geographic dynamics, and especially of dynamic fields and processes, as well as in terms of their enhanced potential for analysis and prediction of their behavior in the space and time. As we indicated by different case studies presented in this paper, the advantages Voronoi diagram can offer for the representation and simulation of a dynamic field and process include the following:

- It is a data model that can be easily adapted to the complex configuration of spatial 2D and 3D entities in the space.
- It allows clear definitions of the spatial relations between objects both in 2D and in 3D spaces, essential in turn for the definition of the interactions between neighboring features in the space.
- It offers an adaptive underlying mesh and grid for geosimulation purposes.
- The dynamic and kinetic operations in Voronoi diagram are well adapted for the simulation purposes, which increases the efficiency and the interactivity of the simulation system.

It should be mentioned, however, that despite these advantages, several complexities associated with the diagram limits its application to simulation purposes, the rectification of which would require further investigation. One such limitation is related to the complexity involved in its implementation especially in more complex spatial entities such as line segments and polygons as well as in volumes in 3D space as compared to the simpler implementation of regular rectangular, triangular or hexagonal grids. Another important shortcoming is the compromised robustness of the diagram's implementation in view of some complex and degeneracy cases. Still another limitation of application of the diagram to simulation purposes is related to the hierarchical representation of a geographic dynamics. Although some initial work has been done on the Voronoi hierarchies for points, the generation of the aggregated objects and their related Voronoi diagram at different meso and macro levels need further investigation.

Finally, spatial and temporal resolutions of a dynamic phenomena and the determination of the initial values and boundary conditions are very important for a successful simulation process. These should be considered as some of main challenges involved in the employment of the Voronoi diagram for the representation and simulation of geographic dynamics.

Acknowledgement

The authors would like to thank NSERC and the Canadian GEOIDE Network of Centers of Excellence for funding this research work. They are also thankful for the helpful comments made by the reviewers of the present paper.

References

[1] Goodchild, M.F., Yuan, M., Cova, T.J.: Towards a General Theory of Geographic Representation in GIS. International Journal of Geographic Information Science 21, 239–260 (2007)
[2] Worboys, M., Duckham, M.: GIS: A Computing Perspective, 2nd edn. CRC Press, Boca Raton (2004)
[3] Peuquet, D.: Representaion of Space and Time. Guilford Press, New York (2002)
[4] Goodchild, M.F.: Representing Fields. Core Curriculum in GIScience, National Center for Geographic Information and Analysis (NCGIA), Santa Barbara (1997)
[5] Cova, T.J., Goodchild, M.F.: Extending geographical Representation to Include Fields of Spatial Objects. International Journal of Geographical Information Science 16, 509–532 (2002)

[6] Mark, D.: Spatial Representation: A Cognitive View. In: Maguire, D.J., Goodchild, M.F., Rhind, D.W., Longley, P. (eds.) Geographical Information Systems: Principles and Applications, 2nd edn., vol. 1, pp. 81–89 (1999)

[7] Bivand, R., Lucas, A.: Integrating Models and Geographical Information Systems. In: Openshaw, S., Abrahart, R.J. (eds.) Geocomputation, pp. 331–364. Taylor & Francis, London (1997)

[8] Chapman, L., Thornes, J.E.: The Use of Geographical Information Systems in Climatology and meteorology. J. Progress in Physical Geography 27(3), 313–330 (2003)

[9] Sui, D.Z., Maggio, R.C.: Integrating GIS with Hydrological Modeling: Practices, Problems, and Prospects. J. Computers, Environment and Urban Systems 23, 33–51 (1999)

[10] Valavanis, V.D.: Geographic Information Systems in Oceanography and Fisheries. Taylor & Francis, London (2002)

[11] De Berg, M., Van Kreveld, M., Overmars, M., Schwarzkopf, O.: Computational Geometry: Algorithms and Applications, 2nd edn. Springer, Berlin (2000)

[12] Batty, M.: Cities and Complexities: Understanding Cities with Cellular Automata, Agent-Based Models and Fractals (2007)

[13] Gold, C.M., Condal, A.R.: A Spatial Data Structure Integrating GIS and Simulation in a Marine Environment. Marine Geodesy 18, 213–228 (1995)

[14] Mostafavi, M.A.: Development of a Global Dynamic Data Structure. PhD thesis, Laval University, Canada (2002)

[15] Mostafavi, M.A., Gold, C.M.: Developemnet of a Global spatial Data Structure for Marine Simulation. International Journal of Geographical Information Science 18, 211–227 (2004)

[16] Hashemi, L., Mostafavi, M.A., Pouliot, J., Govrilova, M.: Towards 3D Dynamic Field Simulation within GIS: A 3D Kinetic Data Structure for Representation and Management of Dynamic Fields. International Journal of Geographical Information Science (2009) (in press)

[17] Fritts, M.J., Crowley, W.P., Trease, H.: The Free-Lagrange Method. Lecture Notes in Physics, vol. 238. Springer, Heidelberg (1985)

[18] Benenson, I., Torrens, P.N.: Geosimulation: Automata-Based Modeling of Urban Phenomena. Wiley, Chichester (2004)

[19] Okabe, A., Boots, B., Sughihara, K., Chin, S.N.: Spatial Tessellations: Concepts and Applications of Voronoi Diagrams, 2nd edn. Wiley, Chichester (2001)

[20] Lardin, P.: Le diagramme Voronoi généralisé comme support a la simulation des écoulements d'eau souterraine par différence finis intégrées. MSc Thesis, Laval University (1999)

[21] Hale, D.: Atomic Meshes: From Seismic Imaging to Reservoir Simulation. In: Proceedings of the 8th European Conference on the Mathematics of Oil Recovery (2002)

[22] Campbell, J., Shashkov, M.: A compatible Lagrangian Hydrodynamics Algorithm for Unstructured Grids. Selcuk Journal of Applied Mathematics 4, 53–70 (2003)

[23] Gold, C.M., Angel, P.: Voronoi Hierarchies. In: Raubal, M., Miller, H.J., Frank, A.U., Goodchild, M.F. (eds.) GIScience 2006. LNCS, vol. 4197, pp. 99–111. Springer, Heidelberg (2006)

[24] Longley, P.A., Goodchild, M.F., Maguire, D.J., Rhind, D.W.: Geographical Information Systems: Principles, Techniques, Management and Applications. John Wiley and Sons, USA (2005)

[25] Gold, C.M., Chau, M., Dzieszko, M., Goralski, R.: The Marine GIS - Dynamic GIS in Action. International archives of photogrammetry, remote sensing and spatial information sciences 35, Part, 688–693D (2004)

[26] Sullivan, O.: Exploring Spatial Process Dynamics Using Irregular Cellular Automaton Models. Geographical Analysis 33(1), 1–18 (2001)

[27] She, W., Pang, Y.: Development of Voronoi-Based Cellular Automata - an Integrated Dynamic Model for Geographical Information Systems. International Journal of Geographical Information Science 14(5), 455–474 (2001)

[28] Nara, A., Torrens, P.M.: Inner-City Gentrification Simulation Using Hybrid Models of Cellular Automata and Multi-Agent Systems. In: Proceedings of the Geocomputation 2005 conference. University of Michigan, Michigan (2005)

[29] Barros, J.: Urban Growth in Latin American Cities: Exploring Urban Dynamics Through Agent-Based Simulation. Thèse de doctorat, University of London (2004)

[30] Batty, M.: Agents, Cells and Cities: New Representational Models for Simulating Multi-Scale Urban Dynamics. Environment and Planning A 37(8), 1373–1394 (2005)

[31] Yuan, M., Hornsby, K.S.: Computation and Visualization for the Understanding of Dynamics in Geographic Domains: A Research Agenda. CRC Press Inc., Boca Raton (2008)

[32] Hornsby, K.S., Yuan, M.: Understanding Dynamics of Geographic Domains. CRC Press, Taylor and Francis Groups (2008)

[33] Goodchild, M.F., Glennon, A.: Representation and Computation of Geographic Dynamics. In: Hornsby, K.S., Yuan, M. (eds.) Understanding Dynamics of Geographic Domains, pp. 13–30. CRC Press, Boca Raton (2008)

[34] Clarke, K.C., Hoppen, S., Gaydos, L.: A Self-Modifying Cellular Automaton Model of Historical Urbanization in the San Francisco Bay area. Environment and Planning B 24, 247–261 (1997)

[35] Bekey, J.A., Karplus, W.J., Kosgan, B.Y.: Modeling and Simulation: Theory and Practice, vol. 38. Springer, Heidelberg (2003)

[36] Burrough, P.A., Karssenberg, D., Van Deursen, W.: Environmental Modeling with PCRaster. In: Maguire, D.J., Batty, M., Goodchild, M.F. (eds.) GIS, Spatial Analysis and Modeling, pp. 333–356. ESRI Press, Redlands (2005)

[37] Batty, M.: Approaches to Modeling in GIS: Spatial Representation and Temporal dynamics. In: Maguire, D.J., Batty, M., Goodchild, M.F. (eds.) GIS, Spatial Analysis and Modeling, pp. 41–61. ESRI Press, Redlands (2005)

Author Index

Anton, François   139

Barequet, Gill   54
Bhak, Jong   123

Fujii, Hidenori   109

Gavrilova, Marina L.   166
Gold, Christopher   139

Halperin, Dan   1
Hanniel, Iddo   54
Hashemi Beni, Leila   183
Hins Mallet, Karine   183
Honiden, Shinichi   28
Houle, Michael E.   28

Kim, Chong-Min   123
Kim, Deok-Soo   123
Kim, Jae-Kwan   123

Liu, Xin   166

Mioc, Darka   139
Mostafavi, Mir Abolfazl   183
Moulin, Bernard   139

Ryu, Joonghyun   123

Setter, Ophir   1
Sharir, Micha   1
Sommer, Christian   28
Sugihara, Kokichi   109

Wallgrün, Jan Oliver   76
Wolff, Martin   28
Won, Chung-In   123